Springer Theses

Recognizing Outstanding Ph.D. Research

Aims and Scope

The series "Springer Theses" brings together a selection of the very best Ph.D. theses from around the world and across the physical sciences. Nominated and endorsed by two recognized specialists, each published volume has been selected for its scientific excellence and the high impact of its contents for the pertinent field of research. For greater accessibility to non-specialists, the published versions include an extended introduction, as well as a foreword by the student's supervisor explaining the special relevance of the work for the field. As a whole, the series will provide a valuable resource both for newcomers to the research fields described, and for other scientists seeking detailed background information on special questions. Finally, it provides an accredited documentation of the valuable contributions made by today's younger generation of scientists.

Theses may be nominated for publication in this series by heads of department at internationally leading universities or institutes and should fulfill all of the following criteria

- They must be written in good English.
- The topic should fall within the confines of Chemistry, Physics, Earth Sciences, Engineering and related interdisciplinary fields such as Materials, Nanoscience, Chemical Engineering, Complex Systems and Biophysics.
- The work reported in the thesis must represent a significant scientific advance.
- If the thesis includes previously published material, permission to reproduce this must be gained from the respective copyright holder (a maximum 30% of the thesis should be a verbatim reproduction from the author's previous publications).
- They must have been examined and passed during the 12 months prior to nomination.
- Each thesis should include a foreword by the supervisor outlining the significance of its content.
- The theses should have a clearly defined structure including an introduction accessible to new PhD students and scientists not expert in the relevant field.

Indexed by zbMATH.

Seishiro Ono

Topology of Superconductors

Theory of Symmetry Indicators, Classifications of Superconducting Nodes, and High-Throughput Investigations

Doctoral Thesis accepted by
The University of Tokyo, Tokyo, Japan

Springer

Author
Dr. Seishiro Ono
Interdisciplinary Theoretical
and Mathematical Sciences Program
(iTHEMS)
RIKEN
Saitama, Japan

Supervisor
Prof. Haruki Watanabe
The University of Tokyo
Tokyo, Japan

ISSN 2190-5053 ISSN 2190-5061 (electronic)
Springer Theses
ISBN 978-981-97-7398-5 ISBN 978-981-97-7399-2 (eBook)
https://doi.org/10.1007/978-981-97-7399-2

This Springer imprint is published by the registered company Springer Nature Singapore Pte Ltd.
The registered company address is: 152 Beach Road, #21-01/04 Gateway East, Singapore 189721, Singapore

Supervisor's Foreword

The recent exploration of topological phases of matter, from both theoretical and experimental perspectives, has revealed a rich variety of possible topological phases in real materials. Triggered by the discovery of $\mathbb{Z}_2$ topological insulators, materials manifesting various topological phases, including higher-order topological insulators and Weyl/Dirac semimetals, have been proposed theoretically and identified experimentally. A pivotal challenge in this field of condensed matter physics remains the identification of candidate materials for topological superconductors, holding promising applications in quantum computation through the emergence of Majorana zero modes on their surfaces. Dr. Seishiro Ono's doctoral thesis tackled this problem head-on from a theoretical standpoint. A key technique driving the recent extensive search for candidate topological materials is the method of "symmetry indicators." This approach enables the partial determination of the topology of band insulators or semimetals from the limited information of the irreducible representation of the space group in the high-symmetric momentum points of occupied Bloch states. Dr. Ono's doctoral thesis achieves a significant breakthrough by extending this method to topological superconductors. He put forward this achievement by constructing a valuable database encapsulating the properties of all conceivable superconducting phases in materials on the database. The readers of this thesis will be benefited from his well-organized summary of this topic from the basics to the forefront of current research on the classification of topological superconductors.

Tokyo, Japan
November 2023

Prof. Haruki Watanabe

List of Publications

Significant part of the present thesis is adapted from the following published papers.

Chapter 3

- **Seishiro Ono**, Hoi Chun Po, and Ken Shiozaki,
  $\mathbb{Z}_2$-enriched symmetry indicators for topological superconductors in the 1651 magnetic space groups,
  Physical Review Research **3**, 023086 (2021).

Chapter 4

- **Seishiro Ono** and Ken Shiozaki,
  Symmetry-Based Approach to Superconducting Nodes: Unification of Compatibility Conditions and Gapless Point Classifications,
  Physical Review X **12**, 011021 (2022).

Chapter 5

- Feng Tang*, **Seishiro Ono***, Xiangang Wan, and Haruki Watanabe,
  High-Throughput Investigations of Topological and Nodal superconductors,
  Physical Review Letters **129**, 027001 (2022) [Highlighted as *Editors' Suggestion*].

(*These authors equally contributed to this work)

Acknowledgements

I sincerely thank Professor Haruki Watanabe, my Ph.D. supervisor, for giving me many research opportunities and for his support in all respects. His profound insight brought me many wonderful collaborations we had, and I learned countless things from him, which would be the nuts and bolts of doing research.

I am deeply indebted to Hirokazu Kobayashi, Akishi Matsugatani, Yusuke Nomura, Hoi Chun Po, Ken Shiozaki, Feng Tang, Luka Trifunovic, Xiangang Wan, and Youichi Yanase for their cooperation. It was my great pleasure to have many insightful discussions during the collaborations, and I thank them for kindly explaining topics with which I am unfamiliar.

Many friends and colleagues helped me during my Ph.D. years. I would particularly like to thank Yohei Fuji, Hiroyuki Fujita, Yuan-Ming Lu, Masamitsu Hayashi, Shota Imaki, Kansei Inamura, Hayato Kanno, Kohei Kawabata, Takuya Kawada, Ryohei Kobayashi, Norifumi Matsumoto, Yuki Miyazaki, Kai Murai, Takuya Nomoto, Yugo Onishi, Masaki Oshikawa, Hisanori Oshima, Shuhei Oyama, Guanxiong Qu, Takuma Saito, Shu Sasaki, Masatoshi Sato, Sachiko Seto, Takuya Shimazaki, Shuntaro Sumita, Yasuhiro Tada, Yuya Tanizaki, Shinji Tsuneyuki, Masato Tsutsumi, Atsushi Ueda, Zixia Wei, Masahiko Yamada, Yasushi Yoneta, and Mao Yoshii.

I wish to thank my family for their continuous support and encouragement throughout my life. Looking back, I noticed my parents always gave me tremendous support for whatever I wanted to do. I would not be who I am today if it had not been for them.

Last but not least, my warmest thanks go to my wife, who is my best partner and friend, for her continued support, patience, and encouragement. I am pretty sure that I would not be able to complete this thesis without her.

Contents

Chapter 1
Introduction

We start this thesis by explaining our motivation and summarizing our key ideas and results.

1.1 Background

Superconductivity is a central topic in condensed matter physics, and it has been studied over the past century since its first discovery. One of the main reasons why people have been fascinated by superconductivity is that, although electrons usually repel each other, some kinds of attractive interactions between electrons create pairs of electrons. Bardeen-Cooper-Schrieffer (BCS) theory succeeded in explaining "elementary" superconductivity in which phonons mediate the attractive interaction between electrons [1]. However, since uranium-based superconductors and cuprate superconductors were discovered in the 1980s [2, 3], various superconductors that cannot be described by the BCS theory have been found. The ultimate goal of studies on such unconventional superconductors is to understand their mechanisms. Transformation properties of Cooper pairs under spatial symmetries, dubbed *pairing symmetry*, are closely related to how electrons pair up. The pairing symmetry is one of the most accessible pieces of information about the mechanism of a superconductor, and it is classified into an irreducible representation of point groups [4–8].

Pairing symmetries realized in actual experiments are often controversial [9–34]. Superconducting nodes that are gapless domains of the quasiparticle energy spectrum in momentum space provide essential information to identify pairing symmetries. Superconducting nodes can be detected experimentally by measuring the power-law behavior of specific heat and magnetic penetration depth as a function of temperature. Therefore, intensive studies on the relationship between pairing symmetries and superconducting nodes have been conducted [35–48]. In particular, it turns out that multi-band (orbital) effects and nonsymmorphic space group symmetries lead to novel superconducting nodes.

S. Ono, *Topology of Superconductors*, Springer Theses,
https://doi.org/10.1007/978-981-97-7399-2_1

Meanwhile, the past decades have witnessed the tremendous development of topological phases of matter that are not necessarily superconducting [49–122]. Spatial symmetries also play a crucial role in studies of topological insulators. On the one hand, they enrich the diversity of topological phases by protecting new topology. The richness of topological crystalline phases is exemplified by mirror Chern insulators [123, 124], topological insulators with hourglass surface states [125], and higher-order topological insulators [100–108]. More recently, theoretical frameworks for classifying topological crystalline phases have been developed [109, 126–132], although these abstract works do not provide topological invariants that are necessary for the diagnosis of topology of materials. On the other hand, representations of spatial symmetries furnished by Bloch wave functions can diagnose topological phases. The prototypical examples are the Fu-Kane parity criterion for $\mathbb{Z}_2$-topological insulators [133] and formulas for Chern number based on inversion and rotation eigenvalues [134–136]. Furthermore, this idea has been generalized to all nonmagnetic and magnetic space groups [98, 99, 137–141], which are nowadays known as *symmetry-indicator theory* and *topological quantum chemistry*. This scheme is not complete but reasonably practical. In fact, symmetry-indicator theory and topological quantum chemistry have enabled large-scale material investigations [142–147].

The topological point of view developed in topological insulators gives rise to a new paradigm of superconductivity. Topological superconductors have attracted much attention owing to the emergence of Majorana fermions on their surfaces [148–150]. Such Majorana fermions could be used for topological quantum computation [151, 152]. Recent intensive studies reveal that unconventional superconductors are likely to exhibit topological superconductivity [153–172]. For example, odd-parity superconductors are promising candidates for topological superconductors [173, 174]. Topology is also intertwined with superconducting nodes [175–187]. Topological invariants defined on subregions of momentum space protect superconducting nodes. For instance, the nonzero Chern number on a sphere indicates the existence of Weyl nodes inside of the sphere. Importantly, possible topological and nodal superconducting phases depend on pairing symmetries [82–84, 86, 87, 188, 189].

These results imply that a comprehensive study on topology of superconductors could help to identify pairing symmetries of realistic materials. Experimental studies first observe characteristics of superconductivity, such as the existence/absence of superconducting nodes and surface states originating from topology. Based on these observations, experimentalists then discuss which pairing symmetry is realized in a material. This step requires knowledge about the topological and nodal nature of each possible pairing symmetry for a given normal phase. This is what theoretical studies can provide.

As an example, let us discuss a spinful system with time-reversal, inversion, and twofold rotation symmetries. There are four possible pairing symmetries in this symmetry setting, but here we focus on two odd-parity pairings denoted by A_u and B_u.[1] For these pairings, half of the number of Fermi surfaces encircling time-reversal

[1] The difference between these two pairing symmetries is the transformation under twofold rotation.

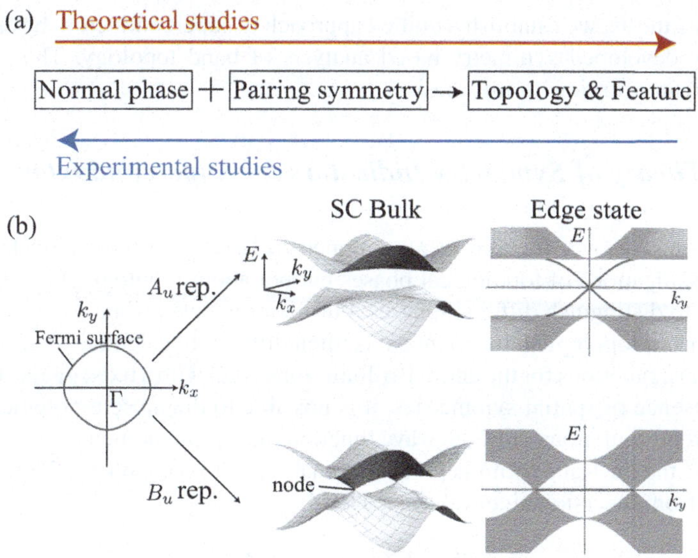

Fig. 1.1 Relations between topology and pairing symmetries. a Flowchart of theoretical and experimental works on topology of superconductors. For a given band structure in the normal phase and a pairing symmetry, theoretical studies predict topological and nodal nature of superconductivity. On the other hand, experimental studies observe phenomena originating from topology or superconducting nodes, and then we identify an actual pairing symmetry of a material. **b** Illustration of relationship among normal-phase band structures, pairing symmetries, and topology in the presence of time-reversal, inversion, and twofold rotation symmetries. For a normal-phase band structure with Fermi surfaces shown in the panel, when A_u pairing symmetry is considered, the system must be a topological superconductor and exhibit helical edge states. In contrast, for B_u pairing symmetry, the system has point nodes protected by the one-dimensional winding number. These results can be understood by symmetry indicators

invariant momenta (TRIMs) correspond to the $\mathbb{Z}_2$ topological invariant for the strong $\mathbb{Z}_2$ topological superconducting phase in two dimension [173, 174]. If the phase has Fermi surfaces shown in Fig. 1.1b, the $\mathbb{Z}_2$ topological invariant is nontrivial according to this criterion. For the A_u case, the strong $\mathbb{Z}_2$ topological superconducting phase can exist, and thus the superconducting phase is topologically nontrivial. In contrast, for the B_u case, the strong $\mathbb{Z}_2$ invariant is always trivial if the system is gapped. This contradiction implies the existence of superconducting nodes, and the system actually has point nodes protected by the one-dimensional winding number.

1.2 Overview

Despite the significant progress in seminal works, a complete understanding of relations among pairing symmetries, topology, and superconducting nodes is still elusive. In this thesis, we establish a unified approach to superconductors by employing a recently developed symmetry-based analysis of band topology. This section is devoted to an overview of this thesis.

1.2.1 Theory of Symmetry Indicators for Superconductors

As mentioned above, for almost all symmetry classes, we have only knowledge about classifications of topological phases but not about topological invariants that enable us to distinguish different phases. Furthermore, even when we have concrete expressions of topological invariants, it is often difficult to calculate them since they require wave functions for the entire Brillouin zone (BZ) with a fixed gauge. However, in the presence of spatial symmetries, it is possible to diagnose topological phases by using representations of Bloch wave functions at certain momenta. A prototypical example is the Fu-Kane formula [133] that enables us to compute the $\mathbb{Z}_2$ topological invariant from inversion eigenvalues at TRIMs:

$$(-1)^{\nu} = \prod_{\boldsymbol{k}\in \mathrm{TRIMs}} \prod_{m\in \mathrm{occ}} \eta_{2m,\boldsymbol{k}} = \prod_{\boldsymbol{k}\in \mathrm{TRIMs}} (-1)^{n_{\boldsymbol{k}}^{-}/2}, \tag{1.1}$$

where $\eta_{2m,\boldsymbol{k}} = \pm 1$ is an inversion eigenvalue of the $2m$-th Bloch wave function at TRIM $\boldsymbol{k}$. Also, $n_{\boldsymbol{k}}^{-}$ is the number of occupied states with negative parities at $\boldsymbol{k}$. This formula tells us whether $\mathbb{Z}_2$ topology of a time-reversal and inversion symmetric insulator is trivial only from four points in two dimensions or eight points in three dimensions [see Fig. 1.2]. It should be mentioned that the authors of Ref. [133] derived this formula from the original expression of the $\mathbb{Z}_2$ invariant; namely, they first had the invariant without inversion, and then they rewrote the invariant in terms of inversion eigenvalues.

Along this line, the authors of Refs. [98] and [99] have developed a symmetry-based approach to band topology independently, which are known as symmetry-indicator theory and topological quantum chemistry. Remarkably, they do not require *a priori* knowledge of topological invariants. In the following, we first review key ideas of symmetry-indicator theory, and then we explain two difficulties in the generalization of the theory to superconductors.

Key ideas of symmetry-indicator theory

The theory consists of two main steps: (i) count of irreducible representations in band structures and (ii) comparison against irreducible representations of trivial insulators.

Count of irreducible representations in band structures.—Let us consider a space group. Some of its elements transform momentum $\boldsymbol{k}$ into other momenta, while others keep $\boldsymbol{k}$ invariant up to reciprocal lattice vectors. A subgroup of the space group is defined at each momentum, whose elements do not change the momentum. This subgroup is called *little group*.[2] In this scheme, a band structure is characterized

[2] We will provide a precise definition of little groups for magnetic space groups in Chap. 2.

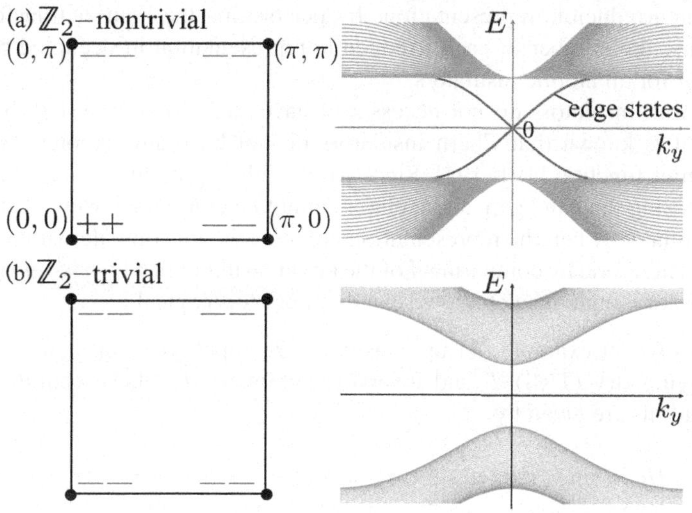

Fig. 1.2 Relation between inversion parities and $\mathbb{Z}_2$ topology. a An inversion parity configuration resulting in $\mathbb{Z}_2$ nontrivial phase. One can see the edge states originating from nontrivial topology. **b** An inversion parity configuration resulting in $\mathbb{Z}_2$ trivial phase. There does not exist any edge state

by the numbers of irreducible representations (a generalized notion of symmetry eigenvalues) of little groups appearing in occupied bands at various momenta, and they form a list

$$\vec{n} = \left(n_{\boldsymbol{k}_1}^{\alpha_1}, n_{\boldsymbol{k}_1}^{\alpha_2}, \cdots, n_{\boldsymbol{k}_2}^{\beta_1}, n_{\boldsymbol{k}_2}^{\beta_2}, \cdots, n_{\boldsymbol{k}_N}^{\gamma_N}\right)^T \in \mathbb{Z}^D, \tag{1.2}$$

where $n_{\boldsymbol{k}}^{\alpha}$ is the count of an irreducible representation of the little group at $\boldsymbol{k}$, and D is a dimension of the list $\vec{n}$. Here, we consider the list of integers $\vec{n}$ as an element of an abelian group $\mathbb{Z}^D$.[3] This process can be understood as a map from band structures to elements of an abelian group.

Comparison against irreducible representations of trivial insulators.—To diagnose topological phases, we must understand what states are trivial since the "nontrivial phase" is a relative concept. In the context of band topology, we define trivial states by product states in terms of real-space orbitals, i.e., all electrons sit at lattice sites without any hopping [98, 99].[4] In this thesis, we refer to such product states as *atomic insulators*, following Ref. [98]. They are insulting by definition, and we can

[3] These integers are always positive; thus, the list is actually an element of a monoid. However, we can promote a monoid to an abelian group. Here, this trick is used.

[4] In the literature, we say that the system is trivial when there exists a deformation preserving all symmetries and a band gap into a system whose ground state is a product state (or equivalently, they are connected to each other by a local unitary transformation) [190].

obtain their irreducible representations by performing the Fourier transformation. By counting the number of each irreducible representation in occupied bands, we construct $\vec{n}$ for all atomic insulators.

In general, insulators do not necessarily have such an atomic description. For example, it is known that Chern insulators do not have any exponentially localized Wannier function [191–194]. Since irreducible representations cannot change without closing the band gap, an insulator cannot be continuously connected to any atomic insulators when the representation contents do not coincide with any atomic insulator. Therefore, by comparing $\vec{n}$ of the target insulator in which we are interested with those for atomic insulators, we can diagnose topological phases.

Example.—As an example, let us consider a $\mathbb{Z}_2$ topological insulator with time-reversal symmetry (TRS) $\mathcal{T}$ and inversion symmetry I, whose Hamiltonian and representations are given by

$$H_{\boldsymbol{k}} = \sin k_x \tau_x \sigma_x + \sin k_y \tau_x \sigma_y + (1 - \cos k_x - \cos k_y)\tau_z \sigma_0, \tag{1.3}$$

$$U(\mathcal{T}) = i\tau_0 \sigma_y, \tag{1.4}$$

$$U(I) = \tau_z \sigma_0, \tag{1.5}$$

where τ_i and σ_i $(i = x, y, z)$ are Pauli matrices. After computing inversion parities at TRIMs, we find

$$\vec{n} = (n_\Gamma^+, n_\Gamma^-, n_X^+, n_X^-, n_Y^+, n_Y^-, n_M^+, n_M^-) = (2, 0, 0, 2, 0, 2, 0, 2)^T \in \mathbb{Z}^8. \tag{1.6}$$

Next, we construct $\vec{n}$ for atomic insulators. When we place an electron with s-orbital at $\boldsymbol{r} = (0, 0)$ in the unit cell, we obtain $\vec{n} = (2, 0, 2, 0, 2, 0, 2, 0)^T$ for this atomic insulator. After performing the same procedures for other atomic insulators, we find that $\vec{n}$ of any atomic insulator can be expressed by the linear combinations of the following five lists with integer coefficients:

$$\vec{a}_1 = (2, 0, 2, 0, 2, 0, 2, 0)^T, \tag{1.7}$$

$$\vec{a}_2 = (0, 2, 0, 2, 0, 2, 0, 2)^T, \tag{1.8}$$

$$\vec{a}_3 = (2, 0, 0, 2, 2, 0, 0, 2)^T, \tag{1.9}$$

$$\vec{a}_4 = (2, 0, 2, 0, 0, 2, 0, 2)^T, \tag{1.10}$$

$$\vec{a}_5 = (2, 0, 0, 2, 0, 2, 2, 0)^T. \tag{1.11}$$

We can consider an abelian group of atomic insulators, and the abelian group $\mathbb{Z}^5$ is spanned by $\{\vec{a}_i\}_{i=1}^5$. If $\vec{n}$ of our target insulator can be expanded by the above vectors with integer coefficients, the insulator is an element of the abelian group. Conversely, when it cannot be done, the insulator is not an element of the abelian group, i.e., the insulator cannot be continuously connected to any atomic insulator. In this example, we have

$$\vec{n} = -\frac{1}{2}\vec{a}_1 + \frac{1}{2}\vec{a}_3 + \frac{1}{2}\vec{a}_4 + \frac{1}{2}\vec{a}_5. \tag{1.12}$$

The fractional coefficients indicate that the system is not an element of the abelian group of atomic insulators. Indeed, this is the same result as the Fu-Kane formula.

Generalization of symmetry-indicator theory to superconductors

Having in mind the success of the symmetry-indicator theory for insulators, it is natural to ask if the theory could be generalized to superconductors to facilitate the discovery of topological superconductors. Since Bogoliubov quasiparticle spectra can be regarded as band structures, it is tempting to think that we can directly apply symmetry-indicator theory to Bogoliubov quasiparticle spectra. However, when we do that, we face the following two problems: (i) symmetry-indicator theory for insulators does not cover well-known symmetry-indicator formulas using $\mathbb{Z}_2$-valued Pfaffian invariants; (ii) it cannot detect one-dimensional topological phases. To generalize the theory to superconductors, we resolve these problems.[5]

Incorporating Pfaffian invariants into symmetry-indicator theory.—In the original theory, we characterize systems by a set of integers, the numbers of occupied states with irreducible representations at all inequivalent high-symmetry momenta. These integer-valued quantities are exactly zero-dimensional topological invariants for normal conducting systems [68, 69]. According to topological classifications for Altland-Zirnbauer (AZ) symmetry classes [68, 69], a $\mathbb{Z}_2$-valued quantity dubbed *Pfaffian invariant* exists in the presence of particle-hole symmetry (PHS). As a result, we characterize superconductors by not only $\mathbb{Z}$-valued topological invariants but also $\mathbb{Z}_2$-valued ones defined for irreducible representations of little groups. In fact, several symmetry-indicator formulas using $\mathbb{Z}_2$-valued Pfaffian invariants exist, for example,

$$\mathrm{Ch} = P_{(0,0)} + P_{(\pi,0)} + P_{(0,\pi)} + P_{(\pi,\pi)} \mod 2, \tag{1.13}$$

where $\mathrm{Ch} \in \mathbb{Z}$ is the Chern number and $P_{(k_x,k_y)}$ is the Pfaffian invariant at (k_x, k_y). Furthermore, the $\mathbb{Z}_2$-nature requires a novel mathematical treatment as shown in Chap. 3.

To characterize a superconductor by these zero-dimensional topological invariants, we have to know which topological invariants are defined for irreducible representations of little groups at momenta. This is achieved by identifying emergent Altland-Zirnbauer (EAZ) classes for the irreducible representations at various high-symmetry momenta. In superconducting phases, the EAZ classes generally depend on pairing symmetries for a given symmetry group. After identifying EAZ classes for irreducible representations at each momentum, we compute the zero-dimensional topological invariants and form a list like Eq. (1.2).

[5] Originally, Refs. [195, 196] discussed how to resolve them for certain symmetry classes. In this thesis, we provide a general framework to overcome these difficulties for arbitrary symmetry classes.

Let us demonstrate how to determine EAZ classes through a simple symmetry setting, space group $P2/m$ with B_g pairing. In this symmetry class, there are TRS $\mathcal{T}$ ($\mathcal{T}^2 = -1$), PHS $\mathcal{C}$ ($\mathcal{C}^2 = +1$), the two-fold rotation C_2^y along the y-axis, and inversion I. The symmetry representations $\{U(g)\}_g$ satisfy

$$U(C_2^y)U(\mathcal{T}) - U(\mathcal{T})U^*(C_2^y) = U(I)U(\mathcal{T}) - U(\mathcal{T})U^*(I) = 0, \tag{1.14}$$

$$U(C_2^y)U(\mathcal{C}) + U(\mathcal{C})U^*(C_2^y) = U(I)U(\mathcal{C}) - U(\mathcal{C})U^*(I) = 0. \tag{1.15}$$

Let $H_{\boldsymbol{k}}$ and $\psi_{m\boldsymbol{k}}$ be a superconducting Hamiltonian and its m-th eigenvector with the quasiparticle energy $E_{m\boldsymbol{k}}$, i.e., $H_{\boldsymbol{k}}\psi_{m\boldsymbol{k}} = E_{m\boldsymbol{k}}\psi_{m\boldsymbol{k}}$. Every point is invariant under the combined symmetries $I\mathcal{C}$ and $I\mathcal{T}$, and thus $U(I\mathcal{C})\psi_{m\boldsymbol{k}}^*$ and $U(I\mathcal{T})\psi_{m\boldsymbol{k}}^*$ are also eigenvectors of $H_{\boldsymbol{k}}$ with the energies $-E_{m\boldsymbol{k}}$ and $E_{m\boldsymbol{k}}$, respectively.

We begin by focusing on a generic momentum $\boldsymbol{k}$ in a two-dimensional plane at $k_y = 0$ or π, which is invariant under the mirror symmetry $M_y = IC_2^y$. In this plane, since $[U(M_y), H_{\boldsymbol{k}}] = 0$, we can simultaneously diagonalize $H_{\boldsymbol{k}}$ and $U(M_y)$. Here, $\psi_{m\boldsymbol{k}}$ denotes the simultaneous eigenvector with eigenenergy $E_{m\boldsymbol{k}}$ and mirror eigenvalue $\xi_{m\boldsymbol{k}} = \pm i$. Then, $U(I\mathcal{C})\psi_{m\boldsymbol{k}}^*$ and $U(I\mathcal{T})\psi_{m\boldsymbol{k}}^*$ have the mirror eigenvalues $\xi_{m\boldsymbol{k}}$ and $-\xi_{m\boldsymbol{k}}$, respectively. This suggests that the combined symmetry $I\mathcal{T}$ interchanges the mirror sector while $I\mathcal{C}$ leaves them unchanged. As a result, we find that the EAZ class for each mirror-eigenvalue sector is class D, as shown in Fig. 1.3a. In the same way, we find that EAZ class of each rotation-eigenvalue sector in a rotation axis is class D as shown in Fig. 1.3b. Thus, the Pfaffian invariants $P_{\boldsymbol{k}}^{\pm i}$ are defined for both cases.

Introducing a difference structure.—As we will discuss in Sect. 3.2.2, the representation contents of the inversion symmetric Kitaev chain [56] coincide with those of an atomic insulator. More generally, since symmetry-indicator theory for insulators is silent for one-dimensional systems, the naive application of symmetry-indicator theory to superconductors fails to capture one-dimensional topological superconductors. This failure can be attributed to the fact that the numbers of irreducible representations of little groups are not zero-dimensional topological invariants of superconductors. Refs. [195, 196] revealed that instead of directly treating our target system in which we are interested, introducing a reference Hamiltonian played a crucial role in diagnosing topological phases of odd-parity superconductors. Indeed, the difference between the numbers of an irreducible representation of our target system and the reference is a $\mathbb{Z}$-valued zero-dimensional topological invariant defined for the irreducible representation. This "difference" structure naturally arises when we revisit trivial superconductors, as discussed in Ref. [198].

1.2.2 Classification of Superconducting Nodes

We now move on to classifications of superconducting nodes, whose importance is stressed in Sect. 1.1. There exist various intriguing nodes that are not covered by

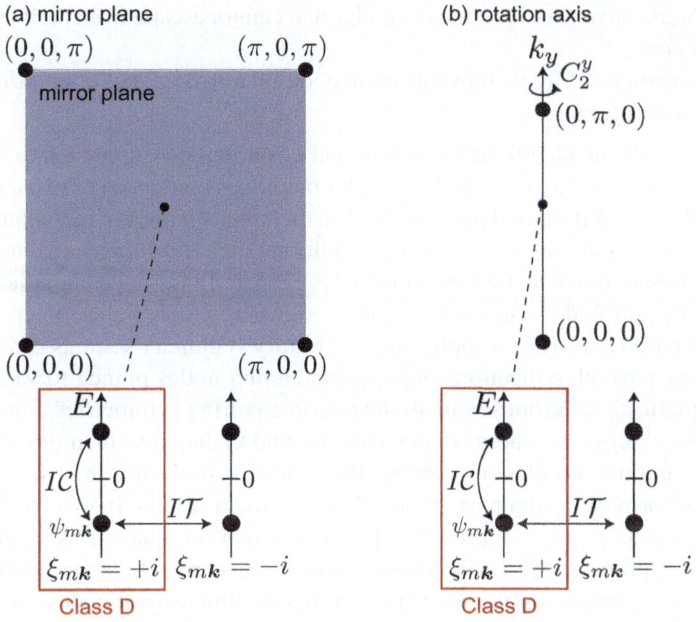

Fig. 1.3 Emergent Altland-Zirnbauer classes for the example. Here, we consider the mirror plane (k_z-k_x plane) (**a**) and the rotation axis (k_y-axis) (**b**), of which every momentum is invariant under IT and IC. Hamiltonians on the mirror plane or the rotation axis are decomposed into two eigenvalue sectors. These two sectors are interchanged by IT but not by IC. Therefore, the effective internal symmetry class for each sector can be understood as class D, and two $\mathbb{Z}_2$-valued Pfaffian invariants are defined at every point in the mirror plane and the rotation axis (except for inversion symmetric momenta). Reprinted from Ref. [197] by The Author(s) licensed under CC BY 4.0

seminal works [4–8]. While several approaches have been proposed to overcome the insufficiency of the previous studies, these proposed theories predominantly focus on several simple symmetry settings, such as generic points or mirror planes.[6] However, actual materials are usually more intricate, and we often encounter complicated symmetry settings, such as a rotation/screw axis in a glide plane. To establish a comprehensive theory, we divide the problem into two parts:

[6] There are three approaches:(i) the group-theoretical analysis of representations of the Cooper pair wave functions in the presence of inversion symmetry [36, 39, 47, 48, 199], (ii) homotopy theory [71, 175–179, 181–185, 200–203], and (iii) $k \cdot p$ model analysis. The first approach informs us of pairing symmetries that force superconducting order parameters to vanish on the mirror plane [36, 48]. Then, Fermi surfaces located on the mirror planes result in line nodes. The second approach discusses topological charges that protect nodes at generic points whose possible symmetries are chiral symmetry or the product of inversion and TRS/PHS. The third approach discusses symmetry-allowed mass terms and dispersion in $k \cdot p$ models [180, 204–209]. More recently, exhaustive $k \cdot p$ Hamiltonians have been introduced for the normal conducting phase in all magnetic space groups [210–213].

(I) Comprehensive classifications of nodes that cannot escape from high-symmetric momenta;
(II) Classifications of freely movable point nodes in mirror planes or point/line nodes on generic momenta.

In this thesis, we tackle the first problem and develop a new approach to superconducting nodes pinned to arbitrary lines.[7] Our method is composed of two techniques, compatibility conditions and point-node classifications on lines in momentum space. Compatibility conditions are necessary conditions for zero-dimensional topological invariants to satisfy when a superconductor is gapped. Point-node classifications tell us whether point nodes can exist or not on a line in which we are interested. Our symmetry-based approach is applicable to arbitrary symmetry settings and forms the basis of extensive classifications of superconducting nodes pinned to various lines in all magnetic space groups with all the possible pairing symmetries. Furthermore, our framework offers a scheme to detect nodes and to diagnose their dimensionality in actual materials, which complements the symmetry-indicator theory.

Here, we provide a summary of our ideas and main results. In this thesis, we use the term *gapless point* to represent a point at which the quasiparticle energy gap closes. It should be noted that gapless points on lines are not always genuine point nodes. In fact, gapless points can be part of higher-dimensional nodes, as discussed below.

Compatibility conditions

As discussed in Sect. 1.2.1, zero-dimensional topological invariants are defined for irreducible representations of the little group at each momentum. For gapped regions, these zero-dimensional topological invariants at distinct momenta are not completely independent and subject to symmetry constraints called *compatibility conditions*. More concretely, common zero-dimensional topological invariants must be the same [see Fig. 1.4a]. This is because, when momenta are regarded as deformation parameters, the zero-dimensional topological invariants cannot change without closing the gap. Inversely, if the zero-dimensional topological invariants are different between two points (namely, some compatibility conditions are violated), there must exist gapless points on this line as illustrated in Fig. 1.4b. While violations of compatibility conditions ensure the existence of gapless points on the line, they do not restrict the global forms of superconducting nodes in momentum space. In the following, we discuss how to understand the dimensionality of a gapless point originating from violations of compatibility conditions.

Let us imagine a gapless point pinned to a line in two-dimensional momentum space (see the upper panel of Fig. 1.4c). The gapless point divides the line into two regions where the zero-dimensional topological invariants differ. We further focus on neighborhoods of these two regions (see the lower panel of Fig. 1.4c). Then, let us consider the case where compatibility conditions exist between these two regions and their surrounding regions. These compatibility conditions ensure that the

[7] Although we focus on Problem (I), we discuss a possible generalization of our framework to answer Problem (II) in Chap. 4.

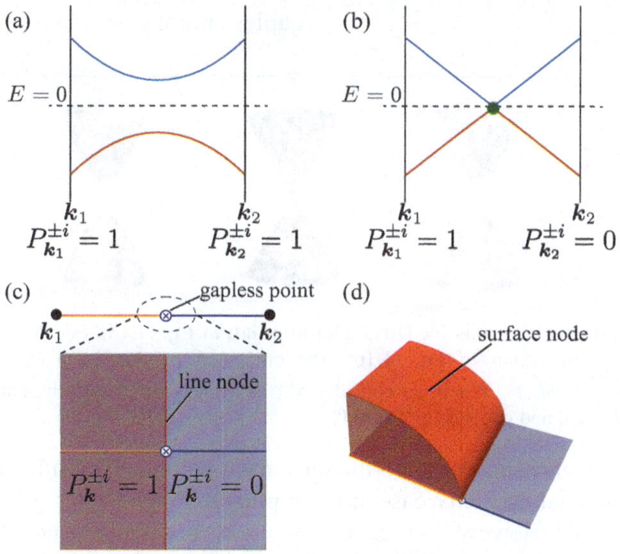

Fig. 1.4 Compatibility conditions and diagnosis of dimensionality. a, b Quasiparticle energy bands on a line connecting two momenta $\boldsymbol{k}_1$ and $\boldsymbol{k}_2$. While compatibility conditions are satisfied on the line (**a**), those are violated in (**b**). **c** [upper] Two regions of the line connecting $\boldsymbol{k}_1$ and $\boldsymbol{k}_2$ in a plane. [lower] a nodal line in the plane. Here, the red and blue regions denote the two domains to which different values of the Pfaffian invariants are assigned. **d** Surface node in three dimensions. If compatibility conditions are present between the plane and its adjacent areas outside the plane, the regions depicted in (**c**) extend outwards in three dimensions. Reprinted from Ref. [197] by The Author(s) licensed under CC BY 4.0

neighborhood domains also divided into two regions that have different topological invariants. Therefore, the boundary line of these neighborhoods is also gapless (see the lower panel of Fig. 1.4c). Surface nodes arise from similar discussions in three dimensions (see Fig. 1.4d).

Here, we discuss the case for space group $P2/m$ with B_g pairing as an example, which is introduced in Sect. 1.2.1. We begin by considering a mirror plane. We pick two generic momenta, denoted by $\boldsymbol{k}_1$ and $\boldsymbol{k}_2$, in the mirror plane. Suppose that Pfaffian invariants at $\boldsymbol{k}_1$ and $\boldsymbol{k}_2$ are different, for example, $P^{\pm i}_{\boldsymbol{k}_1} = 1$ and $P^{\pm i}_{\boldsymbol{k}_2} = 0$. In such a case, as discussed above, a gapless point is pinned to the line between $\boldsymbol{k}_1$ and $\boldsymbol{k}_2$. A compatibility condition is that $P^{\pm i}_{\boldsymbol{k}}$ must be the same for gapped regions in the mirror plane. The current situation is exactly the same as Fig. 1.4c, and thus, the gapless point on the line must be part of a line node.

Next, we consider gapless points on a rotation axis in three-dimensional momentum space of this space group. The discussion for gapless points on rotation axes is completely different from the case of mirror planes. Again, let us consider that there exist gapless points originating from violated compatibility conditions. Furthermore, we focus on neighboring momenta of the rotation axis, which are generic momenta. There are no compatibility conditions between a point on the rotation axis

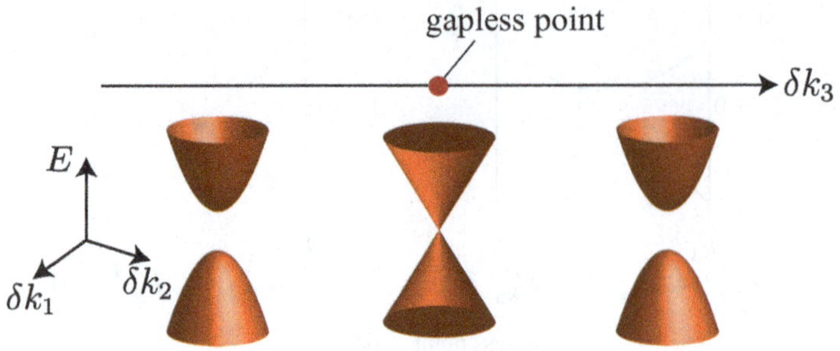

Fig. 1.5 Two-dimensional massive Dirac Hamiltonian in Eq. (1.16). For the two-dimensional massive Dirac Hamiltonian in Eq. (1.16), the energy dispersion is given by $E_{(\delta k_1,\delta k_2)} = \sqrt{(\delta k_1)^2 + (\delta k_2)^2 + (\delta k_3)^2}$. When we consider δk_3 as a mass parameter, the energy dispersion is gapped for $\delta k_3 \neq 0$ and gapless for $\delta k_3 = 0$

and generic momenta since the zero-dimensional topological classification at generic momentum is trivial. In such a case, it is tempting to think that the gapless point is a genuine point node. However, this expectation is not necessarily true. The absence of compatibility conditions guarantees the absence of line and surface nodes protected by zero-dimensional topological invariants, but it does not prohibit the presence of a line node protected by one-dimensional topological invariants such as Berry phase and winding number. Indeed, the gapless point on the rotation axis is not a genuine point node but part of a line node. This result can be understood from the following discussion on point-node classifications.

Point-node classifications on lines

As discussed above, when compatibility conditions between a line and its neighborhood are absent, we need to distinguish between a genuine point node or a line node protected by one-dimensional topological invariants.[8] This is accomplished by classifying two-dimensional massive Dirac Hamiltonians on the line. Let us consider

$$H_{(\delta k_1,\delta k_2)} = \delta k_1 \gamma_1 + \delta k_2 \gamma_2 + \delta k_3 \gamma_0, \tag{1.16}$$

$$\gamma_i \gamma_j + \gamma_j \gamma_i = 2\delta_{ij} \mathbb{1}, \tag{1.17}$$

where δk_1 and δk_2 are displacements from the gapless point in the directions perpendicular to the line, and δk_3 is a displacement along the line [see Fig. 1.5]. Here, we can consider $\delta k_{i=1,2}$ as a momentum parameter and δk_3 as a mass parameter. The energy dispersion is $E_{(\delta k_1,\delta k_2)} = \sqrt{(\delta k_1)^2 + (\delta k_2)^2 + (\delta k_3)^2}$ for the Dirac Hamiltonian, and thus the Hamiltonian is gapped except for $(\delta k_1, \delta k_2, \delta k_3) = (0, 0, 0)$.

When the classification of the Dirac Hamiltonians is nontrivial for a line, stable point nodes can exist on the line. In contrast, if the classification is trivial, there are no stable point nodes on the line. The results can be grouped into three distinct categories:

[8] Surface nodes are protected by zero-dimensional topological charges, and thus they should be understood by compatibility conditions.

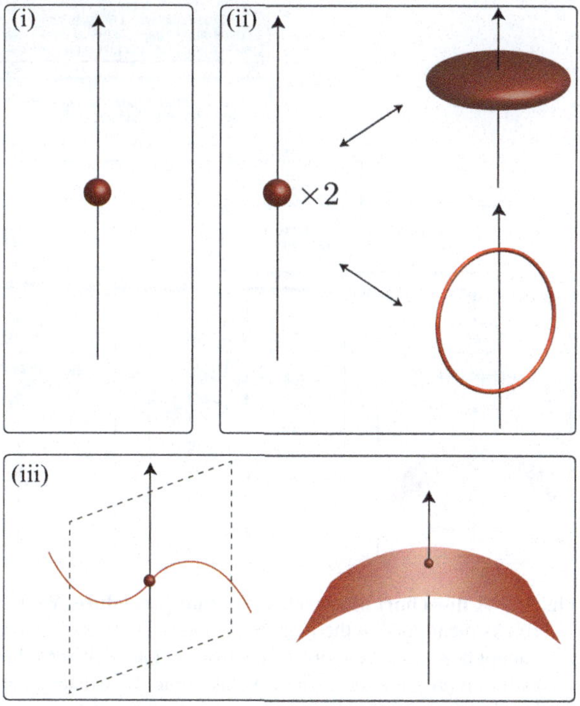

Fig. 1.6 Three types of gapless points on the line. (i) A genuine point node. (ii) A point node composed of multiple gapless points. In fact, such a point node originates from a shrinkable loop or surface node. Without any fine-tuning, a loop or surface node exists since there are no reasons why multiple gapless points are at the same position. (iii) Part of a line or surface node. Reprinted from Ref. [197] by The Author(s) licensed under CC BY 4.0

(i) the classification is nontrivial, and a gapless point on the line is a point node [Fig. 1.6(i)],
(ii) the classification is nontrivial, and a point node is composed of multiple gapless points [Fig. 1.6(ii)],
(iii) the classification is trivial, and thus, if a gapless point exists on the line, it is part of line or surface nodes [Fig. 1.6(iii)].

It should be remarked that for Case (ii), symmetries do not force the multiple gapless points to be in the same positions. According to the classification of the Dirac Hamiltonians, each of the gapless points is not a point node when they are at different positions. In such a case, it is natural to consider that they form a loop or surface node.

A systematic classifications of gapless points in three dimensions
Combining compatibility conditions with point-node classifications, we summarize our classification scheme in the form of a flowchart shown in Fig. 1.7. To carry out our classification systematically, we decompose momentum space into subspaces—points, line segments, polygons, and polyhedrons [see Fig. 4.1 as an example].

We start with a gapless point protected by zero-dimensional topological invariants for irreducible representations of a given magnetic space group. Importantly, this single gapless point cannot be split due to symmetries and has a symmetry-allowed minimum number of gapless states. Here, we call such a gapless point *generator*. We first ask if there exist compatibility conditions between the line and adjacent polygons.

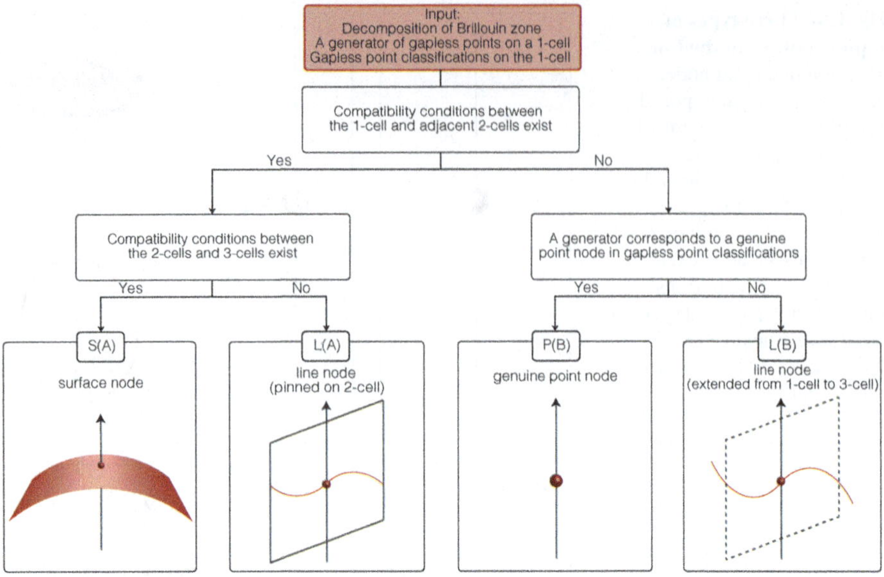

Fig. 1.7 **A flowchart of our classification procedure**. We focus on a generator on a line (called 1-cell). As mentioned in the main text, *generator* refers to the minimum number of gapless points that cannot be split since it originates from a change of zero-dimensional topological invariants for irreducible representations. First, we examine the existence of compatibility conditions between the 1-cell and adjacent polygons (2-cells). If they exist, we further ask if compatibility conditions are between the 2-cells and inner of their adjacent polyhedrons (3-cells). The existence/absence of the compatibility conditions implies that the generator is part of a surface node/line node pinned to the 2-cells [see S(A)/L(A)]. When the compatibility conditions between the 1-cell and 2-cells do not exist, we compare the generator with the results of point-node classifications on 1-cells. When the generator corresponds to case (i) of the point-node classifications, it is a genuine point node [see P(B)]. If the generator does not coincide with a genuine point node, i.e., the point-node classification result is the case (ii) or (iii), we conclude that the generator is part of a line node [see L(B)]. Reprinted from Ref. [197] by The Author(s) licensed under CC BY 4.0

(A) *Compatibility conditions between the line and adjacent polygons exist.*
If they exist, as discussed above, there are two regions with different topological invariants, which results in a gapless line on each of the polygons. Then, by examining compatibility conditions, we further check if the gapless region is extended to adjacent polyhedrons (see Fig. 4.8). If so, the gapless point is part of a surface node [S(A) in Fig. 1.7]. Otherwise, it is part of a line node pinned to the polygons [L(A) in Fig. 1.7].

(B) *Any compatibility condition between the line and adjacent polygons does not exist.*
On the other hand, if there are no compatibility conditions between the line and adjacent polygons, we ask whether the generator is a genuine point node by comparing the generator with our results of point-node classifications. If the generator corresponds to Case (i) of the point-node classification, the generator is a genuine point node [P(B) in Fig. 1.7]. Otherwise, we conclude that the

generator is part of a line node protected by one-dimensional topological invariants [L(B) in Fig. 1.7]. It should be mentioned that surface nodes are always detected by compatibility conditions since they are protected by zero-dimensional topological invariants.

1.2.3 High-Throughput Material Investigation

The theories of symmetry indicators and topological quantum chemistry enable efficient topological material discoveries. In fact, by integrating them into first-principles calculations, large-scale topological material investigations have been conducted [142–147], which have already helped experimentalists discover new topological materials [214]. As mentioned in Sect. 1.2.1, the symmetry-indicator theory has been generalized to superconductors, and thus, one might think that building such a database for superconductors would facilitate the search for new candidates of topological/nodal superconductors. However, this task is unfeasible since pairing symmetries of materials are usually unknown.

Sidestepping this difficulty, we apply an alternative approach: By using the symmetry-indicator theory, we establish a comprehensive database summarizing the topological and nodal nature of *all possible pairing symmetries* for each material. Even if the actual pairing symmetry has not been identified yet, the prediction for each of possible pairing symmetries is still helpful in elucidating the pairing mechanisms of superconductivity in unconventional superconductors by comparing our results with experimental results.

Weak-pairing assumption

In our diagnostic scheme, we make an assumption that the energy scale of the superconducting gap is far smaller than that of normal conducting phases, which is known as *weak-pairing assumption* [150]. This assumption enables us to deduce the quasiparticle energy spectrum from the normal-phase band structure (see Fig. 1.8). In such a case, we can compute zero-dimensional topological invariants in the superconducting phase from the numbers of irreducible representations in the normal phase. For example, the Pfaffian invariant is obtained from the number of occupied bands modulo two.

Applications to materials

Diagnostic scheme.—We are ready to integrate our theories into density functional theory (DFT) calculations. The flowchart of our diagnostic scheme is shown in Fig. 1.9. We perform DFT calculations to obtain irreducible representations at high-symmetry momenta in the normal conducting phase.[9] Then, based on the weak

[9] While all inequivalent high-symmetry momenta must be considered, one can also include more additional momenta in the computation. Computational cost increases, but it sometimes provides more fruitful information about band topology.

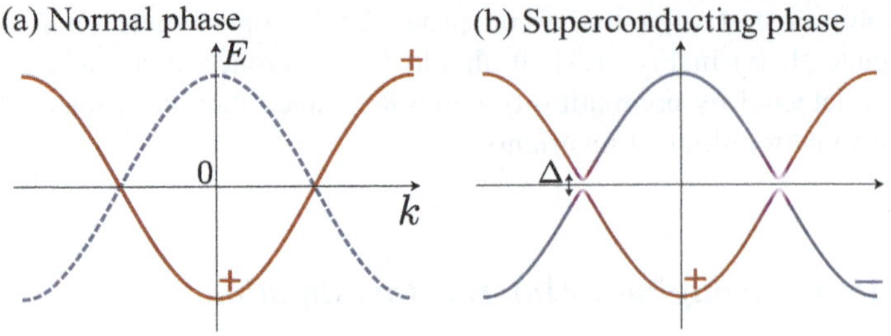

Fig. 1.8 Weak-pairing assumption. **a** A band structure of a one-dimensional metal with a single electron band. Here, we assume that the Bloch wave function at $k = 0$ and π have positive parities. To compare it with the Bogoliubov quasiparticle spectrum in the superconducting phase, here we show a hole band denoted by the dashed blue line. **b** The energy spectrum of the superconducting Hamiltonian with an odd-parity pairing in the regime where the weak-pairing assumption is valid. One can see that the spectrum and parities can be deduced from the normal-phase ones

pairing assumption, we can compute zero-dimensional topological invariants from the irreducible representations for an assumed pairing symmetry. Once we have zero-dimensional topological invariants at high-symmetry momenta, we can diagnose topological and nodal nature in the superconducting phase. First, we check whether all compatibility conditions are satisfied or not. Suppose that there exists a violated compatibility condition between two high-symmetry momenta. In this case, the material has a gapless point on the line connecting them. From our classifications of superconducting nodes, we can identify the positions and dimensionality of the nodes. On the other hand, when all compatibility conditions are satisfied, we compare the set of zero-dimensional topological invariants with those of vacuum states. If the zero-dimensional topological invariants do not coincide with any vacuum state, the system must be topological. If they do, the system does not have any nontrivial topology detected by the symmetry indicators.

Material Investigation.— We apply our diagnostic scheme to 27,208 out of more than 180,000 materials in Inorganic Crystal Structure Database (ICSD) [215] by assuming all of them would be superconducting at low temperatures.[10] For each material, we predict the topological and nodal properties of the superconducting phase by considering various possible pairing symmetries. Our results could reduce the candidates of pairing symmetries when they are compared with experimental results.

According to results of DFT calculations, 10,822 of them have Fermi surfaces aside from their size, and we focus on them here.[11] When we restrict our computations to nontrivial pairing symmetries corresponding to one-dimensional representations, 88.9% of computed materials are identified as topological or nodal superconductors,

[10] Many materials, whose superconductivity has not been experimentally confirmed, are included in our computations. However, here we assume that they would become superconductors when the temperature decreases sufficiently. In fact, the superconducting transition at around eight milli-Kelvin has recently been observed in $YbRh_2Si_2$ [216].

[11] All computed materials are listed in the database Database of Topological and Nodal Superconductors (https://tnsc.nju.edu.cn).

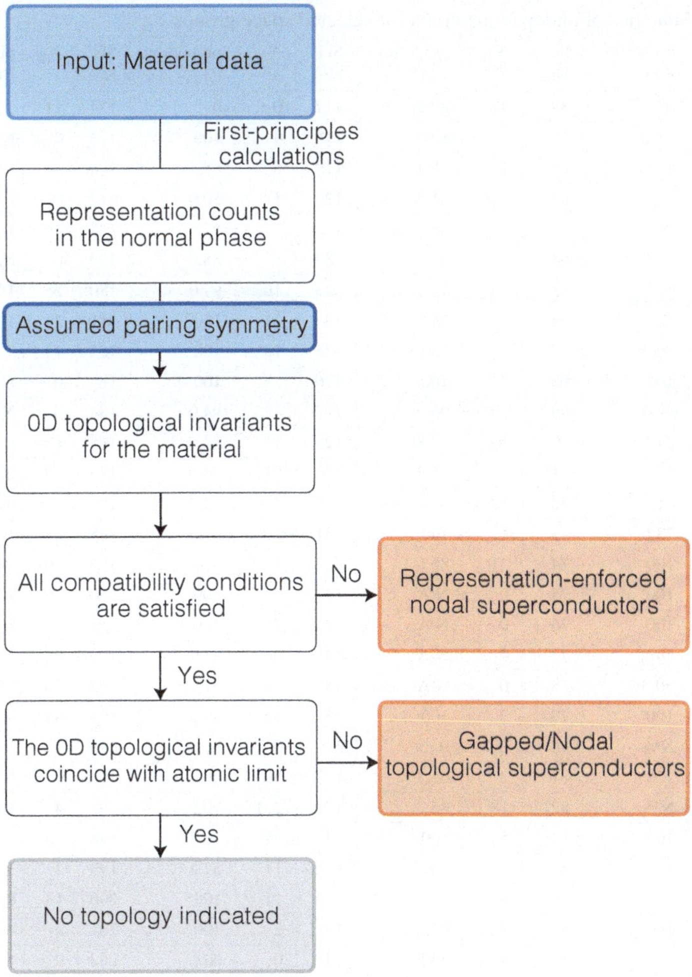

Fig. 1.9 Flowchart of our diagnostic scheme. For a given material and an assumed pairing symmetry, we diagnose the topological and nodal superconducting nature based on the theories in Sects. 1.2.1 and 1.2.2. This figure is a simplified version of Fig. 5.1

irrespective of the assumed nontrivial pairing symmetries. This observation agrees with the empirical fact that a superconductor tends to be nontrivial if its pairing symmetry is nontrivial. From a material synthesis point of view, it is worthwhile to discuss the percentage for each space group (see Table 1.1). Remarkably, 22 space groups have high percentages. Hence, by studying materials in these space groups, we have a better chance to find new topological and nodal superconductors.

We also provide a subroutine Topological Supercon (http://toposupercon.t.u-tokyo.ac.jp/tms), which can efficiently diagnose the topological and nodal superconducting nature of any material synthesized in the future. Also, this program can handle various conditions not covered by our database.

Table 1.1 Statistics of undoped materials for selected space groups

SG No.	N_M	proportion (%)	SG No.	N_M	proportion (%)	SG No.	N_M	proportion (%)	SG No.	N_M	proportion (%)
1	0	0	59	41	87.2	117	0	0	174	11	91.7
2	125	91.2	60	17	89.5	118	4	100	175	0	N/A
3	1	100	61	19	86.4	119	6	85.7	176	87	95.6
4	21	77.8	62	556	81.4	120	1	50.0	177	0	N/A
5	20	90.9	63	386	85.2	121	7	87.5	178	1	100
6	2	100	64	47	82.5	122	8	88.9	179	0	N/A
7	10	100	65	56	93.3	123	163	97.0	180	26	100
8	9	56.2	66	0	N/A	124	11	100	181	4	100
9	3	60.0	67	3	100	125	10	100	182	12	92.3
10	6	100	68	3	100	126	2	100	183	0	N/A
11	87	92.6	69	29	96.7	127	125	94.0	184	0	N/A
12	286	87.2	70	36	87.8	128	11	84.6	185	13	86.7
13	9	100	71	123	95.3	129	213	93.0	186	69	69.7
14	116	81.7	72	17	81.0	130	13	72.2	187	128	97.0
15	64	73.6	73	1	100	131	10	62.5	188	2	100
16	1	100	74	71	86.6	132	1	100	189	167	94.4
17	2	100	75	0	N/A	133	2	100	190	19	100
18	3	100	76	0	N/A	134	7	100	191	227	100
19	0	N/A	77	0	N/A	135	6	85.7	192	0	N/A
20	2	50.0	78	0	N/A	136	64	71.1	193	266	97.4
21	6	100	79	2	100	137	12	85.7	194	834	92.2
22	0	N/A	80	0	0	138	0	N/A	195	1	100
23	1	100	81	2	66.7	139	505	90.7	196	0	N/A
24	0	N/A	82	18	85.7	140	231	90.9	197	4	100
25	12	100	83	5	100	141	99	76.7	198	56	86.2
26	4	66.7	84	4	100	142	11	52.4	199	2	33.3
27	0	N/A	85	4	66.7	143	11	100	200	13	81.2
28	3	100	86	30	93.8	144	2	100	201	2	40.0
29	2	100	87	49	84.5	145	0	N/A	202	0	N/A
30	0	N/A	88	11	57.9	146	10	83.3	203	0	N/A
31	5	41.7	89	0	N/A	147	11	100	204	17	41.5
32	0	N/A	90	1	100	148	135	67.5	205	28	50.0
33	3	37.5	91	0	N/A	149	1	50.0	206	16	94.1
34	0	0	92	15	71.4	150	5	83.3	207	0	N/A
35	0	N/A	93	0	N/A	151	1	100	208	5	100
36	14	56.0	94	1	100	152	6	100	209	0	N/A
37	0	N/A	95	0	N/A	153	1	100	210	1	100
38	20	95.2	96	0	N/A	154	2	100	211	3	100
39	2	66.7	97	1	100	155	6	100	212	6	60.0
40	2	66.7	98	0	N/A	156	8	72.7	213	11	78.6

(continued)

Table 1.1 (continued)

SG No.	N_M	proportion (%)	SG No.	N_M	proportion (%)	SG No.	N_M	proportion (%)	SG No.	N_M	proportion (%)
41	6	66.7	99	0	N/A	157	4	80.0	214	11	100
42	16	88.9	100	1	100	158	1	100	215	13	100
43	9	40.9	101	0	N/A	159	8	88.9	216	272	97.1
44	21	87.5	102	3	75.0	160	51	87.9	217	37	100
45	0	N/A	103	7	100	161	7	100	218	4	100
46	5	100	104	1	100	162	18	100	219	0	N/A
47	6	100	105	1	100	163	15	78.9	220	77	95.1
48	0	N/A	106	0	N/A	164	225	95.7	221	621	99.8
49	0	N/A	107	36	87.8	165	13	100	222	0	N/A
50	3	60.0	108	4	80.0	166	298	95.5	223	109	77.3
51	44	95.7	109	9	52.9	167	96	94.1	224	10	100
52	2	40.0	110	0	N/A	168	0	N/A	225	812	96.4
53	3	100	111	0	N/A	169	0	N/A	226	24	53.3
54	8	100	112	1	100	170	0	0	227	242	88.6
55	76	82.6	113	8	88.9	171	0	N/A	228	0	N/A
56	3	75.0	114	3	50.0	172	0	N/A	229	165	99.4
57	14	82.4	115	6	85.7	173	14	93.3	230	3	100
58	32	54.2	116	2	66.7						

For each space group, we count the number of materials classified into nodal superconductors or topological superconductors for all pairings except for the trivial irreducible representation of the point group, and the number is denoted by N_M. We also show the percentages of such materials to the total number of materials in each space group. Reproduced from Supplementary Table II of Ref. [217]

Table 1.2 Results for BaPtSb

PG	Pairing symmetry	Result	Nodes and topology
D_{3h}	A_2''	Nodal line	K-H
	A_2'	Nodal line	Γ-K, M-K, K-H
	A_1''	Topological	$(0, 1, 0, 0, 5, 5) \in (\mathbb{Z}_2)^4 \times (\mathbb{Z}_6)^2$
	A_1'	N/A	–

The first column is the point group of space group $P\bar{6}m2$, and symbols in the second column mean assumed pairing symmetries. Definitions of these representations are included in Supplemental Materials of Ref. [217]. The fourth column indicates nodal and topological nature of each pairing. The last column represents an entry of symmetry indicators for A_1'' and the paths where compatibility conditions are broken for A_2'' and A_2' representations or the entry of symmetry indicators for A_1'' representation. Adapted from Table 1 of Ref. [217]

Examples.— Here, we pick BaPtSb from our database as an example, whose constituent elements are the same families as those in unconventional superconductor candidates CaPtAs and SrPtAs [31, 158]. Superconductivity was recently reported [218]. The results of our symmetry-based analysis are summarized in Table 1.2.

The space group is $P\bar{6}m2$ (No. 187), in which inversion symmetry is absent [215]. The point group of this space group is D_{3h}, and there are four one-dimensional single-valued representations: A_2'', A_2', A_1'' and A_1'.[12] For A_1' representation, our approach is silent.[13] We find that all compatibility conditions hold for the A_1'' pairing. Thus, we can apply the symmetry-indicator theory to this case. The entry of the symmetry indicator is $(0, 1, 0, 0, 5, 5) \in (\mathbb{Z}_2)^4 \times (\mathbb{Z}_6)^2$, which indicates that the mirror Chern number on the $k_z = \pi$ equals one modulo three while that on the $k_z = 0$ plane is zero modulo three. Conversely, the system is a nodal superconductor for A_2'' or A_2' pairing. There exist various violated compatibility conditions along the lines listed in Table 1.2. Given that all lines lie within the Γ-A-H-K or M-L-H-K planes, any violations lead to line nodes on these planes [197].

1.3 Organization of the Thesis

The remainder of this thesis is organized as follows.

- In Chap. 2, we provide basics of (magnetic) space-group symmetries in band theory. In particular, we show how such spatial symmetries are incorporated into band theory, which representation theory describes. Then, we review compatibility relations and atomic insulators, which are key ingredients in the theory of symmetry indicators.
- In Chap. 3, we first introduce the mean-field Hamiltonian and symmetry representations in the superconducting phases. Then, we present two main problems in a naive application of the original symmetry-indicator theory to superconductors. We also provide their resolutions: (i) incorporating Pfaffian invariants into symmetry-indicator theory for superconductors and (ii) introducing a "difference" structure to the theory.
- In Chap. 4, we formulate our classification theory for superconducting nodes based on *Atiyah-Hirzebruch spectral sequence* [221], which enables us to perform the classifications systematically. Readers who are not interested in the detailed formulation of our theory can skip this chapter.
- In Chap. 5, we show how to use the above two theories for diagnosing topological and nodal superconductivity in realistic materials. In particular, we discuss recently discovered superconductors CaPtAs, BaPtP, and CsV_3Sb_5. These results could be compared with experimental results to narrow down the list of potential pairing symmetries.
- In Chap. 6, we conclude this thesis with an outlook for future studies.

---

[12] There are also two-dimensional irreducible representations, but we focus on one-dimensional ones here.

[13] Our symmetry-based analysis is always silent for trivial pairing symmetries in the presence of TRS. In fact, it is known that there exist various gapped topological phases for such symmetry classes [219, 220].

References

1. Bardeen J, Cooper LN, Schrieffer JR (1957) Microscopic theory of superconductivity. Phys Rev 106:162–164
2. Stewart GR, Fisk Z, Willis JO, Smith JL (1984) Possibility of coexistence of bulk superconductivity and spin fluctuations in UPt_3. Phys Rev Lett 52:679–682
3. Bednorz JG, Müller KA (1986) Possible High T_c Superconductivity in the Ba-La-Cu-O System. Zeitschrift für Physik B Condens Matter 64(2):189–193
4. Volovik GE, Gor'kov LP (1985) Superconducting classes in heavy-fermion systems. Zh Eksp Teor Fiz 88:1412–1428
5. Anderson PW (Oct1984) Structure of "triplet" superconducting energy gaps. Phys Rev B 30:4000–4002
6. Ozaki M, Machida K, Ohmi T (1985) On p-wave pairing superconductivity under cubic symmetry. Progr Theoret Phys 74(2):221–235
7. Ozaki M, Machida K, Ohmi T (1986) On p-wave pairing superconductivity under hexagonal and tetragonal symmetries. Progr Theoret Phys 75(2):442–444
8. Manfred S, Kazuo U (1991) Phenomenological theory of unconventional superconductivity. Rev Mod Phys 63:239–311
9. Ishida K, Mukuda H, Kitaoka Y, Asayama K, Mao ZQ, Mori Y, Maeno Y (1998) Spin-triplet superconductivity in Sr_2RuO_4 identified by ^{17}O Knight shift. Nature 396(6712):658–660
10. Luke GM, Fudamoto Y, Kojima KM, Larkin MI, Merrin J, Nachumi B, Uemura YJ, Maeno Y, Mao ZQ, Mori Y, Nakamura H, Sigrist M (1998) Time-reversal symmetry-breaking superconductivity in Sr_2RuO_4. Nature 394(6693):558–561
11. Shingo Y, Tomohiro K, Yoshiteru M (2013) First-order superconducting transition of Sr_2RuO_4. Phys Rev Lett 110:077003
12. Shunichiro K, Akira K, Toshiro S, Daisuke S, Shingo Y, Yoshiteru M, Kenichi T, Kazushige M (2014) Sharp magnetization jump at the first-order superconducting transition in Sr_2RuO_4. Phys Rev B 90:220502(R)
13. Hassinger E, Bourgeois-Hope P, Taniguchi H, René de Cotret S, Grissonnanche G, Anwar MS, Maeno Y, Doiron-Leyraud N, Taillefer L (2017) Vertical line nodes in the superconducting gap structure of Sr_2RuO_4. Phys Rev X 7:011032
14. Yasui Y, Lahabi K, Anwar MS, Nakamura Y, Yonezawa S, Terashima T, Aarts J, Maeno Y (2017) Little-Parks oscillations with half-quantum fluxoid features in Sr_2RuO_4 microrings. Phys Rev B 96:180507(R)
15. Kittaka S, Nakamura S, Sakakibara T, Kikugawa N, Terashima T, Uji S, Sokolov DA, Mackenzie AP, Irie K, Tsutsumi Y, Suzuki K, Machida K (2018) Searching for Gap Zeros in Sr_2RuO_4 via field-angle-dependent specific-heat measurement. J Phys Soc Jap 87(9):093703
16. Pustogow A, Luo Y, Chronister A, Su YS, Sokolov DA, Jerzembeck F, Mackenzie AP, Hicks CW, Raghu KS, Bauer ED, Brown SE (2019) Constraints on the superconducting order parameter in Sr_2RuO_4 from oxygen-17 nuclear magnetic resonance. Nature 574(7776):72–75
17. Kashiwaya S, Saitoh K, Kashiwaya H, Koyanagi M, Sato M, Yada K, Tanaka Y, Maeno Y (2019) Time-reversal invariant superconductivity of Sr_2RuO_4 revealed by Josephson effects. Phys Rev B 100:094530 Sep
18. Ishida K, Manago M, Kinjo K, Maeno Y (2020) Reduction of the ^{17}O knight shift in the superconducting state and the heat-up effect by NMR pulses on Sr_2RuO_4. J Phys Soc Jap 89(3):034712
19. Kivelson SA, Yuan AC, Ramshaw B, Thomale R (2020) A proposal for reconciling diverse experiments on the superconducting state in Sr_2RuO_4. NPJ Quant Mater 5(1):43
20. Chronister A, Pustogow A, Kikugawa N, Sokolov DA, Jerzembeck F, Hicks CW, Mackenzie AP, Bauer ED, Brown SE (2020) Evidence for even parity unconventional superconductivity in Sr_2RuO_4

21. Sheng R, Chris E, Qing-Ping D, Yuji F, Tristin M, Saha Shanta R, Liu I-L, Zic M, Kim H, Paglione J, Butch NP (2019) Nearly ferromagnetic spin-triplet superconductivity. Science 365(6454):684–687
22. Ishizuka J, Sumita S, Daido A, Yanase Y (2019) Insulator-metal transition and topological superconductivity in UTe_2 from a first-principles calculation. Phys Rev Lett 123:217001 Nov
23. Xu Y, Sheng Yutao, Yang Y (2019) Quasi-two-dimensional fermi surfaces and unitary spin-triplet pairing in the heavy fermion superconductor UTe_2. Phys Rev Lett 123:217002 Nov
24. Metz T, Bae S, Ran S, Liu I-L, Eo YS, Fuhrman WT, Agterberg DF, Anlage SM, Butch NP, Paglione J (2019) Point-node gap structure of the spin-triplet superconductor UTe_2. Phys Rev B 100:220504(R)
25. Jiao L, Howard, S Ran S, Wang Z, Rodriguez JO, Sigrist M, Wang Z, Butch NP, Madhavan V (2020) Chiral superconductivity in heavy-fermion metal UTe_2. Nature 579(7800):523–527
26. Shunichiro K, Yusei S, Toshiro S, Ai N, Dexin L, Yoshiya H, Fuminori H, Dai A, Kazushige M (2020) Orientation of point nodes and nonunitary triplet pairing tuned by the easy-axis magnetization in UTe_2. Phys Rev Res 2:032014
27. Bae S, Kim H, Ran S, Eo YS, Liu I-L, Fuhrman W, Paglione J, Butch NP, Anlage SM (2020) Anomalous normal fluid response in a chiral superconductor UTe_2
28. Hayes IM, Wei DS, Metz T, Zhang J, Eo YS, Ran S, Saha SR, Collini J, Butch NP, Agterberg DF, Kapitulnik A, Paglione J (2020) Weyl Superconductivity in UTe_2
29. Ishizuka J, Yanase Y (2021) Periodic Anderson model for magnetism and superconductivity in UTe_2. Phys. Rev. B 103:094504 Mar
30. Goryo J, Fischer MH, Sigrist M (2012) Possible pairing symmetries in SrPtAs with a local lack of inversion center. Phys Rev B 86:100507(R) Sep
31. Biswas PK, Luetkens H, Neupert T, Stürzer T, Baines C, Pascua G, Schnyder AP, Fischer MH, Goryo J, Lees MR, Maeter H, Brückner F, Klauss H-H, Nicklas M, Baker PJ, Hillier AD, Sigrist M, Amato A, Johrendt D (2013) Evidence for superconductivity with broken time-reversal symmetry in locally noncentrosymmetric SrPtAs. Phys Rev B 87:180503(R) May
32. Fischer MH, Neupert T, Platt C, Schnyder AP, Hanke W, Goryo J, Thomale R, Sigrist M (2014) Chiral d-wave superconductivity in SrPtAs. Phys Rev B 89:020509(R) Jan
33. Matano K, Arima K, Maeda S, Nishikubo Y, Kudo K, Nohara M, Zheng G (2014) Spin-singlet superconductivity with a full gap in locally noncentrosymmetric SrPtAs. Phys Rev B 89:140504(R)
34. Fischer MH, Goryo J (2015) Symmetry and Gap Classification of Non-Symmorphic SrPtAs. J Phys Soc Jap 84(5):054705
35. Norman MR (1995) Odd parity and line nodes in heavy-fermion superconductors. Phys Rev B 52:15093–15094 Dec
36. Micklitz T, Norman MR (2009) Odd parity and line nodes in nonsymmorphic superconductors. Phys Rev B 80:100506(R) Sep
37. Nomoto T, Ikeda H (2016) Exotic multigap structure in UPt_3 unveiled by a first-principles analysis. Phys Rev Lett 117:217002 Nov
38. Yanase Y (2016) Nonsymmorphic Weyl superconductivity in UPt_3 based on E_{2u} representation. Phys Rev B 94:174502 Nov
39. Micklitz T, Norman MR (2017) Nodal lines and nodal loops in nonsymmorphic odd-parity superconductors. Phys Rev B 95:024508 Jan
40. Brydon PMR, Wang L, Weinert M, Agterberg DF (2016) Pairing of $j = 3/2$ fermions in half-heusler superconductors. Phys Rev Lett 116:177001
41. Agterberg DF, Brydon PMR, Timm C (2017) Bogoliubov fermi surfaces in superconductors with broken time-reversal symmetry. Phys Rev Lett 118:127001 Mar
42. Timm C, Schnyder AP, Agterberg DF, Brydon PMR (2017) Inflated nodes and surface states in superconducting Half-Heusler compounds. Phys Rev B 96:094526 Sep
43. Savary L, Ruhman J, Venderbos JWF, Fu L, Lee PA (2017) Superconductivity in three-dimensional spin-orbit coupled semimetals. Phys Rev B 96:214514

44. Kim H, Wang K, Nakajima Y, Hu R, Ziemak S, Syers P, Wang L, Hodovanets H, Denlinger JD, Brydon PMR, Agterberg DF, Tanatar MA, Prozorov R, Paglione J (2018) Beyond triplet: unconventional superconductivity in a spin-3/2 topological semimetal. Sci Adv 4(4):eaao4513
45. Boettcher I, Herbut IF (2018) Unconventional superconductivity in luttinger semimetals: theory of complex tensor order and the emergence of the uniaxial nematic state. Phys Rev Lett 120:057002 Feb
46. Venderbos JWF, Savary L, Ruhman J, Lee PA, Fu L (2018) Pairing states of spin-$\frac{3}{2}$ fermions: symmetry-enforced topological gap functions. Phys Rev X 8:011029
47. Nomoto T, Ikeda H (2017) Symmetry-protected line nodes in Non-symmorphic magnetic space groups: applications to UCoGe and UPd_2Al_3. J Phys Soc Jap 86(2):023703
48. Sumita S, Yanase Y (2018) Unconventional superconducting gap structure protected by space group symmetry. Phys Rev B 97:134512 Apr
49. Thouless DJ, Kohmoto M, Nightingale MP, den Nijs M (1982) Quantized hall conductance in a two-dimensional periodic potential. Phys Rev Lett 49:405–408 Aug
50. Kohmoto M (1985) Topological invariant and the quantization of the hall conductance. Ann Phys 160(2):343–354
51. Haldane FDM (1988) Model for a quantum hall effect without landau levels: condensed-matter realization of the "parity anomaly". Phys Rev Lett 61:2015–2018 Oct
52. Hatsugai Y (1993) Edge states in the integer quantum hall effect and the Riemann surface of the Bloch function. Phys Rev B 48:11851–11862 Oct
53. Hatsugai Y (1993) Chern number and edge states in the integer quantum hall effect. Phys Rev Lett 71:3697–3700 Nov
54. Michel L, Zak J (1999) Connectivity of energy bands in crystals. Phys Rev B 59:5998–6001 Mar
55. Michel L, Zak J (2000) Elementary energy bands in crystalline solids. Europhys Lett 50(4):519–525
56. Kitaev YuA (2001) Unpaired Majorana fermions in quantum wires. Physics-Uspekhi 44(10S):131–136
57. Murakami S, Nagaosa N, Zhang S-C (2003) Dissipationless quantum spin current at room temperature. Science 301(5638):1348–1351
58. Murakami S, Nagaosa N, Zhang S-C (2004) Spin-hall insulator. Phys Rev Lett 93:156804 Oct
59. Kane CL, Mele EJ (2005) Z_2 topological order and the quantum spin hall effect. Phys Rev Lett 95:146802 Sep
60. Kane CL, Mele EJ (2005) Quantum spin hall effect in graphene. Phys Rev Lett 95:226801 Nov
61. Andrei Bernevig B, Hughes TL, Zhang S-C (2006) Quantum spin hall effect and topological phase transition in hgte quantum wells. Science 314(5806):1757–1761
62. Fu L, Kane CL (2006) Time reversal polarization and a Z_2 adiabatic spin pump. Phys Rev B 74:195312
63. König M, Wiedmann S, Brüne C, Roth A, Buhmann H, Molenkamp LW, Qi X-L, Zhang S-C (2007) Quantum spin hall insulator state in HGTE quantum wells. Science 318(5851):766–770
64. Fu L, Kane CL, Mele EJ (2007) Topological insulators in three dimensions. Phys Rev Lett 98:106803
65. Moore JE, Balents L (2007) Topological invariants of time-reversal-invariant band structures. Phys Rev B 75:121306 Mar
66. Hsieh D, Qian D, Wray L, Xia Y, Hor YS, Cava RJ, Hasan MZ (2008) A topological dirac insulator in a quantum spin hall phase. Nature 452(7190):970–974
67. Schnyder AP, Ryu S, Furusaki A, Ludwig AWW (2008) Classification of topological insulators and superconductors in three spatial dimensions. Phys Rev B 78:195125
68. Kitaev A (2009) Periodic table for topological insulators and superconductors. AIP Conf Proc 1134(1):22–30

69. Ryu S, Schnyder AP, Furusaki A, Ludwig AWW (2010) Topological insulators and superconductors: tenfold way and dimensional hierarchy. New J Phys 12(6):065010
70. Sato M, Fujimoto S (2009) Topological phases of noncentrosymmetric superconductors: edge states, Majorana fermions, and non-Abelian statistics. Phys Rev B 79:094504 Mar
71. Teo JCY, Kane CL (2010) Topological defects and gapless modes in insulators and superconductors. Phys Rev B 82:115120
72. Chen X, Gu Z-C, Wen X-G (2010) Local unitary transformation, long-range quantum entanglement, wave function renormalization, and topological order. Phys Rev B 82:155138 Oct
73. Chen X, Gu Z-C, Wen X-G (2011) Classification of gapped symmetric phases in one-dimensional spin systems. Phys Rev B 83:035107 Jan
74. Chen X, Gu Z-C, Liu Z-X, Wen X-G (2012) Symmetry-protected topological orders in interacting Bosonic systems. Science 338(6114):1604–1606
75. Chen X, Gu Z-C, Liu Z-X, Wen X-G (2013) Symmetry protected topological orders and the group cohomology of their symmetry group. Phys Rev B 87:155114 Apr
76. Hasan MZ, Kane CL (2010) Colloquium: topological insulators. Rev Mod Phys 82:3045–3067 Nov
77. Fu L (2011) Topological crystalline insulators. Phys Rev Lett 106:106802 Mar
78. Pollmann F, Turner AM, Berg E, Oshikawa M (2010) Entanglement spectrum of a topological phase in one dimension. Phys Rev B 81:064439 Feb
79. Wan X, Turner AM, Vishwanath A, Savrasov SY (2011) Topological semimetal and fermi-arc surface states in the electronic structure of pyrochlore iridates. Phys. Rev. B 83:205101 May
80. Pollmann F, Berg E, Turner AM, Oshikawa M (2012) Symmetry protection of topological phases in one-dimensional quantum spin systems. Phys Rev B 85:075125 Feb
81. Slager R-J, Mesaros A, Juričić V, Zaanen J (2013) The space group classification of topological band-insulators. Nat Phys 9(2):98–102
82. Morimoto T, Furusaki A (2013) Topological classification with additional symmetries from Clifford algebras. Phys Rev B 88:125129 Sep
83. Chiu C-K, Yao H, Ryu S (2013) Classification of topological insulators and superconductors in the presence of reflection symmetry. Phys Rev B 88:075142 Aug
84. Shiozaki K, Sato M (2014) Topology of crystalline insulators and superconductors. Phys Rev B 90:165114 Oct
85. Fang C, Chen Y, Kee H-Y, Fu L (2015) Topological nodal line semimetals with and without spin-orbital coupling. Phys Rev B 92:081201
86. Shiozaki K, Sato M, Gomi K (2016) Topology of nonsymmorphic crystalline insulators and superconductors. Phys Rev B 93:195413 May
87. Shiozaki K, Sato M, Gomi K (2017) Topological crystalline materials: general formulation, module structure, and wallpaper groups. Phys Rev B 95:235425 Jun
88. Morimoto T, Nagaosa N (2016) Topological nature of nonlinear optical effects in solids. Science Advances 2(5):e1501524
89. Potter Andrew C, Morimoto Takahiro, Vishwanath Ashvin (2016) Classification of interacting topological floquet phases in one dimension. Phys Rev X 6:041001 Oct
90. Po HC, Fidkowski L, Morimoto T, Potter AC, Vishwanath A (2016) Chiral floquet phases of many-body localized bosons. Phys Rev X 6:041070
91. Murakami S, Hirayama M, Okugawa R, Miyake T (2017) Emergence of topological semimetals in gap closing in semiconductors without inversion symmetry. Sci Adv 3(5):e1602680
92. Po HC, Watanabe H, Zaletel MP, Vishwanath A (2016) Filling-enforced quantum band insulators in spin-orbit coupled crystals. Sci Adv 2(4):e1501782
93. Watanabe H, Po HC, Zaletel MP, Vishwanath A (2016) Filling-enforced gaplessness in band structures of the 230 space groups. Phys Rev Lett 117:096404
94. Chiu C-K, Teo JCY, Schnyder AP, Ryu S (2016) Classification of topological quantum matter with symmetries. Rev Mod Phys 88:035005
95. de Juan F, Grushin AG, Morimoto T, Moore JE (2017) Quantized circular photogalvanic effect in weyl semimetals. Nat Commun 8(1):15995

96. Hirayama M, Okugawa R, Miyake T, Murakami S (2017) Topological dirac nodal lines and surface charges in FCC alkaline earth metals. Nat Commun 8(1):14022
97. Furuya SC, Oshikawa M (2017) Symmetry protection of critical phases and a global anomaly in $1 + 1$ dimensions. Phys Rev Lett 118:021601 Jan
98. Po HC, Vishwanath A, Watanabe H (2017) Symmetry-based indicators of band topology in the 230 space groups. Nat Commun 8(1):50
99. Bradlyn B, Elcoro L, Cano J, Vergniory MG, Wang Z, Felser C, Aroyo MI, Andrei Bernevig B (2017) Topological quantum chemistry. Nature 547(7663):298–305
100. Wladimir A, Benalcazar B, Bernevig A, Hughes TL (2017) Quantized electric multipole insulators. Science 357(6346):61–66
101. Benalcazar WA, Andrei Bernevig B, Hughes TL (2017) Electric multipole moments, topological multipole moment pumping, and chiral hinge states in crystalline insulators. Phys Rev B 96:245115
102. Song Z, Fang Z, Fang C (2017) $(d - 2)$-Dimensional edge states of rotation symmetry protected topological states. Phys Rev Lett 119:246402 Dec
103. Schindler F, Wang Z, Vergniory MG, Cook AM, Murani A, Sengupta S, Kasumov AYu, Deblock R, Jeon S, Drozdov I, Bouchiat H, Guéron S, Yazdani A, Andrei Bernevig B, Neupert T (2018) Higher-order topology in bismuth. Nat Phys 14(9):918–924
104. Schindler F, Cook AM, Vergniory MG, Wang Z, Parkin SSP, Andrei Bernevig B, Neupert T (2018) Higher-order topological insulators. Sci Adv 4(6):aat0346
105. Geier M, Trifunovic L, Hoskam M, Brouwer PW (2018) Second-order topological insulators and superconductors with an order-two crystalline symmetry. Phys Rev B 97:205135 May
106. Khalaf E (2018) Higher-order topological insulators and superconductors protected by inversion symmetry. Phys Rev B 97:205136 May
107. Trifunovic L, Brouwer PW (2019) Higher-order bulk-boundary correspondence for topological crystalline phases. Phys Rev X 9:011012 Jan
108. Fang C, Fu L (2019) New classes of topological crystalline insulators having surface rotation anomaly. Sci Adv 5(12):eaat2374
109. Thorngren R, Else DV (2018) Gauging spatial symmetries and the classification of topological crystalline phases. Phys Rev X 8:011040 Mar
110. Armitage NP, Mele EJ, Vishwanath A (2018) Weyl and dirac semimetals in three-dimensional solids. Rev Mod Phys 90:015001
111. Gong Z, Ashida Y, Kawabata K, Takasan K, Higashikawa S, Ueda M (2018) Topological phases of non-Hermitian systems. Phys Rev X 8:031079 Sep
112. Kawabata K, Bessho T, Sato M (2019) Classification of exceptional points and non-Hermitian topological semimetals. Phys Rev Lett 123:066405 Aug
113. Yokomizo K, Murakami S (2019) Non-bloch band theory of non-Hermitian systems. Phys Rev Lett 123:066404 Aug
114. Kawabata K, Shiozaki K, Ueda M, Sato M (2019) Symmetry and topology in non-Hermitian physics. Phys Rev X 9:041015 Oct
115. Else DV, Po HC, Watanabe H (2019) Fragile topological phases in interacting systems. Phys Rev B 99:125122
116. Okuma N, Kawabata K, Shiozaki K, Sato M (2020) Topological origin of non-Hermitian skin effects. Phys Rev Lett 124:086801 Feb
117. Nakatsuji S, Arita R (2022) Topological magnets: functions based on berry phase and multipoles. Annu Rev Condens Matter Phys 13(1):119–142
118. Zhang K, Yang Z, Fang C (2020) Correspondence between winding numbers and skin modes in non-Hermitian systems. Phys Rev Lett 125:126402 Sep
119. Yang Z, Zhang K, Fang C, Hu J (2020) Non-Hermitian bulk-boundary correspondence and auxiliary generalized Brillouin zone theory. Phys Rev Lett 125:226402 Nov
120. Yu J, Zhang R-X, Song Z-D (2021) Dynamical symmetry indicators for floquet crystals. Nat Commun 12(1):5985
121. Verresen R, Thorngren R, Jones NG, Pollmann F (2021) Gapless topological phases and symmetry-enriched quantum criticality. Phys Rev X 11:041059 Dec

122. Ding K, Fang C, Ma G (2022) Non-Hermitian topology and exceptional-point geometries. Nat Rev Phys 4(12):745–760
123. Tanaka Y, Ren Z, Sato T, Nakayama K, Souma S, Takahashi T, Segawa K, Ando Y (2012) Experimental realization of a topological crystalline insulator in SNTE. Nat Phys 8(11):800–803
124. Hsieh TH, Lin H, Liu J, Duan W, Bansil A, Fu L (2012) Topological crystalline insulators in the SNTE material class. Nat Commun 3(1):982
125. Wang Z, Alexandradinata A, Cava RJ, Andrei Bernevig B (2016) Hourglass fermions. Nature 532(7598):189–194
126. Huang S-J, Song H, Huang Y-P, Hermele M (2017) Building crystalline topological phases from lower-dimensional states. Phys Rev B 96:205106 Nov
127. Song H, Huang S-J, Fu L, Hermele M (2017) Topological phases protected by point group symmetry. Phys Rev X 7:011020 Feb
128. Song Z, Huang S-J, Qi Y, Fang C, Hermele M (2019) Topological states from topological crystals. Sci Adv 5(12):eaax2007
129. Song Z, Fang C, Qi Y (2020) Real-space recipes for general topological crystalline states. Nat Commun 11(1):4197
130. Ken S, Charles ZX, Kiyonori G (2018) Generalized homology and Atiyah-Hirzebruch spectral sequence in crystalline symmetry protected topological phenomena
131. Peng B, Jiang Y, Fang Z, Weng H, Fang C (2022) Topological classification and diagnosis in magnetically ordered electronic materials. Phys Rev B 105:235138 Jun
132. Else DV, Thorngren R (2019) Crystalline topological phases as defect networks. Phys Rev B 99:115116 Mar
133. Fu L, Kane CL (2007) Topological insulators with inversion symmetry. Phys Rev B 76:045302
134. Hughes TL, Prodan E, Andrei Bernevig B (2011) Inversion-symmetric topological insulators. Phys Rev B 83:245132
135. Turner AM, Zhang Y, Mong RSK, Vishwanath A (2012) Quantized response and topology of magnetic insulators with inversion symmetry. Phys Rev B 85:165120
136. Fang C, Gilbert MJ, Andrei Bernevig B (2012) Bulk topological invariants in noninteracting point group symmetric insulators. Phys Rev B 86:115112
137. Kruthoff J, de Boer J, van Wezel J, Kane CL, Slager R-J (2017) Topological classification of crystalline insulators through band structure combinatorics. Phys Rev X 7:041069 Dec
138. Watanabe H, Po HC, Vishwanath A (2018) Structure and topology of band structures in the 1651 magnetic space groups. Sci Adv 4(8):eaat8685
139. Song Z, Zhang T, Fang C (2018) Diagnosis for nonmagnetic topological semimetals in the absence of spin-orbital coupling. Phys Rev X 8:031069 Sep
140. Khalaf E, Po HC, Vishwanath A, Watanabe H (2018) Symmetry indicators and anomalous surface states of topological crystalline insulators. Phys Rev X 8:031070
141. Elcoro L, Wieder BJ, Song Z, Xu Y, Bradlyn B, Andrei Bernevig B (2021) Magnetic topological quantum chemistry. Nat Commun 12(1):5965
142. Tang F, Po HC, Vishwanath A, Wan X (2019) Efficient topological materials discovery using symmetry indicators. Nat Phys 15(5):470–476
143. Tang F, Po HC, Vishwanath A, Wan X (2019) Topological materials discovery by large-order symmetry indicators. Sci Adv 5(3):eaau8725
144. Zhang T, Jiang Y, Song Z, Huang H, He Y, Fang Z, Weng H, Fang C (2019) Catalogue of topological electronic materials. Nature 566(7745):475–479
145. Tang F, Po HC, Vishwanath A, Wan X (2019) Comprehensive search for topological materials using symmetry indicators. Nature 566(7745):486–489
146. Vergniory MG, Elcoro L, Felser C, Regnault N, Andrei Bernevig B, Wang Z (2019) A complete catalogue of high-quality topological materials. Nature 566(7745):480–485
147. Xu, Y Elcoro L, Song Z-D, Wieder BJ, Vergniory MG, Regnault N, Chen Y, Felser C, Andrei Bernevig B (2020) High-throughput calculations of magnetic topological materials. Nature 586(7831):702–707

148. Qi X-L, Zhang S-C (2011) Topological insulators and superconductors. Rev. Mod. Phys. 83:1057–1110 Oct
149. Sato M, Fujimoto S (2016) Majorana fermions and topology in superconductors. J Phys Soc Jpn 85(7):072001
150. Sato M, Ando Y (2017) Topological superconductors: a review. Rep Progr Phys 80(7):076501 May
151. Elliott SR, Franz M (2015) Colloquium: Majorana fermions in nuclear, particle, and solid-state physics. Rev. Mod. Phys. 87:137–163 Feb
152. Lian B, Sun X-Q, Vaezi A, Qi X-L, Zhang S-C (2018) Topological quantum computation based on chiral Majorana fermions. Proc Nat Acad Sci 115(43):10938–10942
153. Fu L (2014) Odd-parity topological superconductor with nematic order: application to $Cu_xBi_2Se_3$. Phys Rev B 90:100509(R) Sep
154. Sasaki S, Kriener M, Segawa K, Yada K, Tanaka Y, Sato M, Ando Y (2011) Topological Superconductivity in $Cu_xBi_2Se_3$. Phys Rev Lett 107:217001
155. Yanase Y, Shiozaki K (2017) Möbius topological superconductivity in UPt_3. Phys. Rev. B 95:224514 Jun
156. Kawakami T, Okamura T, Kobayashi S, Sato M (2018) Topological crystalline materials of $J = 3/2$ electrons: antiperovskites, dirac points, and high winding topological superconductivity. Phys Rev X 8:041026 Nov
157. Daido A, Yoshida T, Yanase Y (2019) Γ_4 topological superconductivity in UCOGE. Phys Rev Lett 122:227001 Jun
158. Shang T, Smidman M, Wang A, Chang L-J, Baines C, Lee MK, Nie ZY, Pang GM, Xie W, Jiang WB, Shi M, Medarde M, Shiroka T, Yuan HQ (2020) Simultaneous nodal superconductivity and time-reversal symmetry breaking in the noncentrosymmetric superconductor CaPtAs. Phys Rev Lett 124:207001 May
159. Hsu Y-T, Cole WS, Zhang R-X, Sau JD (2020) Inversion-protected higher-order topological superconductivity in monolayer WTe_2. Phys Rev Lett 125:097001 Aug
160. Nogaki K, Daido A, Ishizuka J, Yanase Y (2021) Topological crystalline superconductivity in locally noncentrosymmetric $CeRh_2As_2$. Phys Rev Res 3:L032071 Sep
161. Sato M (2009) Topological properties of spin-triplet superconductors and Fermi surface topology in the normal state. Phys Rev B 79:214526 Jun
162. Ueno Y, Yamakage A, Tanaka Y, Sato M (2013) Symmetry-protected Majorana Fermions in topological crystalline superconductors: theory and application to Sr_2RuO_4. Phys Rev Lett 111:087002 Aug
163. Langbehn J, Peng Y, Trifunovic L, von Oppen F, Brouwer PW (2017) Reflection-symmetric second-order topological insulators and superconductors. Phys Rev Lett 119:246401 Dec
164. Cornfeld E, Chapman A (2019) Classification of crystalline topological insulators and superconductors with point group symmetries. Phys Rev B 99:075105 Feb
165. Ahn J, Yang B-J (2020) Higher-order topological superconductivity of spin-polarized fermions. Phys Rev Res 2:012060 Mar
166. Tiwari A, Jahin A, Wang Y (2020) Chiral dirac superconductors: second-order and boundary-obstructed topology. Phys Rev Res 2:043300 Dec
167. Huang S-J, Hsu Y-T (2021) Faithful derivation of symmetry indicators: a case study for topological superconductors with time-reversal and inversion symmetries. Phys Rev Res 3:013243 Mar
168. Cornfeld E, Carmeli S (2021) Tenfold topology of crystals: unified classification of crystalline topological insulators and superconductors. Phys Rev Res 3:013052 Jan
169. Chen Y, Huang S-J, Hsu Y-T, Wei T-C (2022) Topological invariants beyond symmetry indicators: boundary diagnostics for twofold rotationally symmetric superconductors. Phys Rev B 105:094518 Mar
170. Timm C, Bhattacharya A (2021) Symmetry, nodal structure, and Bogoliubov fermi surfaces for nonlocal pairing. Phys Rev B 104:094529 Sep
171. Lo CFB, Po HC, Nevidomskyy AH (2022) Inherited topological superconductivity in two-dimensional dirac semimetals. Phys Rev B 105:104501

172. Yu J, Chen Y-A, Sarma SD (2022) Euler-obstructed cooper pairing: nodal superconductivity and hinge Majorana zero modes. Phys Rev B 105:104515
173. Sato M (2010) Topological odd-parity superconductors. Phys Rev B 81:220504(R) Jun
174. Fu L, Berg E (2010) Odd-parity topological superconductors: theory and application to $Cu_xBi_2Se_3$. Phys Rev Lett 105:097001
175. Yada K, Sato M, Tanaka Y, Yokoyama T (2011) Surface density of states and topological edge states in noncentrosymmetric superconductors. Phys Rev B 83:064505 Feb
176. Sato M, Tanaka Y, Yada K, Yokoyama T (2011) Topology of Andreev bound states with flat dispersion. Phys Rev B 83:224511 Jun
177. Chiu C-K, Schnyder AP (2014) Classification of reflection-symmetry-protected topological semimetals and nodal superconductors. Phys Rev B 90:205136 Nov
178. Matsuura S, Chang P-Y, Schnyder AP, Ryu S (2013) Protected boundary states in gapless topological phases. New J Phys 15(6):065001
179. Kobayashi S, Shiozaki K, Tanaka Y, Sato M (2014) Topological Blount's theorem of odd-parity superconductors. Phys Rev B 90:024516 Jul
180. Yang B-J, Nagaosa N (2014) Classification of stable three-dimensional dirac semimetals with nontrivial topology. Nat Commun 5(1):4898
181. Kobayashi S, Yanase Y, Sato M (2016) Topologically stable gapless phases in nonsymmorphic superconductors. Phys Rev B 94:134512 Oct
182. Bzdušek T, Sigrist M (2017) Robust doubly charged nodal lines and nodal surfaces in centrosymmetric systems. Phys Rev B 96(15):155105
183. Kobayashi S, Sumita S, Yanase Y, Sato M (2018) Symmetry-protected line nodes and Majorana flat bands in nodal crystalline superconductors. Phys Rev B 97:180504(R) May
184. Sumita S, Nomoto T, Shiozaki K, Yanase Y (2019) Classification of topological crystalline superconducting nodes on high-symmetry lines: point nodes, line nodes, and Bogoliubov fermi surfaces. Phys Rev B 99:134513 Apr
185. Kim S, Ryu D-C, Yang B-J (2021) Linking structures of doubly charged nodal surfaces in centrosymmetric superconductors. Phys Rev B 103:224523 Jun
186. Simon S, Geier M, Brouwer PW (2022) Higher-order topological semimetals and nodal superconductors with an order-two crystalline symmetry. Phys Rev B 106:035105 Jul
187. Le C, Yang Z, Cui F, Schnyder AP, Chiu C-K (2022) Generalized fermion doubling theorems: Classification of two-dimensional nodal systems in terms of wallpaper groups. Phys Rev B 106:045126
188. Ken S (2022) The classification of surface states of topological insulators and superconductors with magnetic point group symmetry. Progr Theor Experim Physi 2022(4)
189. Zhang Z, Ren J, Qi Y, Fang C (2022) Topological classification of intrinsic three-dimensional superconductors using anomalous surface construction. Phys Rev B 106:L121108 Sep
190. Zeng B, Chen X, Zhou D, Wen X (2019) Quantum information meets quantum matter: from quantum entanglement to topological phases of many-body systems. Springer, New York March
191. Thonhauser T, Vanderbilt D (2006) Insulator/chern-insulator transition in the haldane model. Phys Rev B 74:235111
192. Soluyanov AA, Vanderbilt D (2011) Wannier representation of $\mathbb{Z}_2$ topological insulators. Phys Rev B 83:035108 Jan
193. Marzari N, Mostofi AA, Yates JR, Souza I, Vanderbilt D (2012) Maximally localized Wannier functions: theory and applications. Rev Mod Phys 84:1419–1475 Oct
194. Alexandradinata A, Höller J (2018) No-go theorem for topological insulators and high-throughput identification of chern insulators. Phys Rev B 98:184305 Nov
195. Skurativska A, Neupert T, Fischer MH (2020) Atomic limit and inversion-symmetry indicators for topological superconductors. Phys Rev Res 2:013064 Jan
196. Ken S (2019) Variants of the symmetry-based indicator
197. Ono S, Ken S (2022) Symmetry-based approach to superconducting nodes: unification of compatibility conditions and gapless point classifications. Phys. Rev. X 12:011021 Feb

198. Ono S, Po HC, Watanabe H (2020) Refined symmetry indicators for topological superconductors in all space groups. Sci Adv 6(18):eaaz8367
199. Micklitz T, Norman MR (2017) Symmetry-enforced line nodes in unconventional superconductors. Phys Rev Lett 118:207001 May
200. Tanaka Y, Sato M, Nagaosa N (2012) Symmetry and topology in superconductors -odd-frequency pairing and edge states-. J Phys Soc Jap 81(1):011013
201. Zhao YX, Wang ZD (2013) Topological classification and stability of fermi surfaces. Phys Rev Lett 110:240404 Jun
202. Bo L, Yada K, Sato M, Tanaka Y (2015) Crossed surface flat bands of Weyl semimetal superconductors. Phys Rev Lett 114:096804 Mar
203. Kobayashi S, Tanaka Y, Sato M (2015) Fragile surface zero-energy flat bands in three-dimensional chiral superconductors. Phys Rev B 92:214514 Dec
204. Young SM, Zaheer S, Teo JCY, Kane CL, Mele EJ, Rappe AM (2012) Dirac semimetal in three dimensions. Phys Rev Lett 108:140405 Apr
205. Phillips M, Aji V (2014) Tunable line node semimetals. Phys Rev B 90:115111 Sep
206. Chen Y, Lu Y-M, Kee H-Y (2015) Topological crystalline metal in orthorhombic perovskite iridates. Nat Commun 6(1):6593
207. Weng H, Fang C, Fang Z, Andrei Bernevig B, Dai X (2015) Weyl semimetal phase in noncentrosymmetric transition-metal monophosphides. Phys Rev X 5:011029
208. Rui Y, Weng H, Fang Z, Dai X, Xiao H (2015) Topological node-line semimetal and dirac semimetal state in antiperovskite cu_3PdN. Phys Rev Lett 115:036807 Jul
209. Wieder BJ, Kim Y, Rappe AM, Kane CL (2016) Double dirac semimetals in three dimensions. Phys Rev Lett 116:186402
210. Jiang Y, Fang Z, Fang C (2021) A $k \cdot p$ effective Hamiltonian generator. Chin Phys Lett 38(7):077104 Jul
211. Yu Z-M, Zhang Z, Liu G-B, Wu W, Li X-P, Zhang R-W, Yang SA, Yao Y (2022) Encyclopedia of emergent particles in three-dimensional crystals. Sci Bull 67(4):375–380
212. Tang F, Wan X (2021) Exhaustive construction of effective models in 1651 magnetic space groups. Phys Rev B 104:085137 Aug
213. Zhan G, Shi M, Yang Z, Zhang H (2021) A Programmable $k \cdot p$ Hamiltonian method and application to magnetic topological insulator Mn_2Bi_4. Chin Phys Lett 38(7):077105 Jul
214. Ryo N, Masaru K, Jiang Z, Kenta K, Takanari T, Xu Z, Lee D, Motoaki H, Masayuki , Tetsuroh S, Zhang P, Lin C, Bareille C, Sakuragi S, Tanaka H, Kunisada S, Kurokawa K, Yaji K, Harasawa A, Kandyba V, Giampietri A, Barinov A, Kim TK, Cacho C, Hashimoto M, Lu D, Shin S, Arita R, Lai K, Sasagawa T, Kondo T (2021) Evidence for a higher-order topological insulator in a three-dimensional material built from van der Waals stacking of bismuth-halide chains. Nat Mater 20:473–479
215. Hellenbrandt M (2004) The inorganic crystal structure database (ICSD)-present and future. Crystallogr Rev 10:17–22
216. Nguyen DH, Sidorenko A, Taupin M, Knebel G, Lapertot G, Schuberth E, Paschen S (2021) Superconductivity in an extreme strange metal. Nat Commun 12(1):4341
217. Tang F, Ono S, Wan X, Watanabe H (2022) High-throughput investigations of topological and nodal superconductors. Phys Rev Lett 129:027001 Jul
218. Kudo K, Saito Y, Takeuchi T, Ayukawa S, Kawamata T, Nakamura S, Koike Y, Nohara M (2018) Superconductivity in BaPtSb with an ordered honeycomb network. J Phys Soc Jap 87(6):063702
219. Qin S, Fang C, Zhang F-C, Hu J (2022) Topological superconductivity in an extended s-wave superconductor and its implication to iron-based superconductors. Phys Rev X 12:011030 Feb
220. Ono S, Ken S, Watanabe H (2022) Classification of time-reversal symmetric topological superconducting phases for conventional pairing symmetries
221. Ken S, Sato M, Gomi K (2022) Atiyah-Hirzebruch spectral sequence in band topology: general formalism and topological invariants for 230 space groups. Phys Rev B 106:165103 Oct

Chapter 2
Review of Symmetry and Representations in Band Theory

As mentioned in Sect. 1.1, crystalline symmetries, such as lattice translations and rotations, play a pivotal role in studies on both superconductivity and topological phases. In this chapter, we review representation theory in band structures and the original symmetry-indicator theory, following discussions in Refs. [1–4]. Although we focus only on normal phases here, most notions introduced in this chapter are common to superconducting phases. In fact, we will use them to introduce representation theory for superconducting phases in the next chapter. Furthermore, the review of symmetry-indicator theory for normal conducting phases would clarify the difference in the theory for superconducting phases, which is discussed in Chap. 3.

In Sect. 2.1, we explain point groups, space groups, and magnetic space groups. In Sect. 2.2, we introduce symmetry representations in band theory, including projective representations and irreducible representations of little (co)groups. In Sect. 2.3, we review various relations among irreducible representations, which originate from symmetries relating to different momenta, nonsymmorphic symmetries, antiunitary symmetries, and band connectivity. In Sects. 2.4 and 2.5, we review two key ideas of symmetry-based approach to band topology: (i) considering a set of all band structures as an abelian group and (ii) defining topologically trivial states by atomic insulators. Once we have the abelian groups of all band insulators and all atomic insulators for a given space group, we can evaluate the mismatch between two abelian groups, which is reviewed in Sect. 2.6. This chapter is a review of key ingredients of symmetry-based analysis of band topology [1, 5], and the notations of this chapter are entirely based on Refs. [2, 3, 6] with some changes.

2.1 Basics of Crystalline Symmetry

In this section, we introduce some basic symmetry groups in solids. Readers interested in more details should consult the classical literature on symmetry and group theory in solids [7–9].

S. Ono, *Topology of Superconductors*, Springer Theses,
https://doi.org/10.1007/978-981-97-7399-2_2

2.1.1 Point Group

A point group is a group of geometric and isometric symmetries that keep at least one point fixed. In particular, point groups discussed in solid-state physics are named as *crystallographic point groups*, but we shortly refer to crystallographic point groups as point groups in this thesis. There are 32 point groups in three spatial dimensions, which are generated by (i) spatial inversion, (ii) n-fold rotations ($n = 2, 3, 4, 6$), and (iii) products of the rotation and inversion.

2.1.2 Space Group

In a crystal, constituent atoms are arrayed with a periodic structure in all directions. Such periodicity is dubbed *discrete translation symmetry*. In the following, we omit *discrete* and simply call it *translation symmetry*. Because of the periodicity, we can define a unit that can span the entire space by translations. Such a unit is referred to as *unit cell*. A unit cell in three dimensions is characterized by a set of vectors $\{\boldsymbol{a}_i\}_{i=1}^{3}$, called *primitive lattice vectors*. Then, the translations form a symmetry group T, called *translation group*, and the group action of $T_{\boldsymbol{R}} \in T$ on $\mathbb{R}^3$ is defined by

$$T_{\boldsymbol{R}}(\boldsymbol{x}) = \boldsymbol{x} + \boldsymbol{R} \in \mathbb{R}^3, \quad \boldsymbol{R} = \sum_{i=1}^{3} n_i \boldsymbol{a}_i, \tag{2.1}$$

where $\{n_i\}_{i=1}^{3}$ are integers.

In addition to the translation symmetries, crystals usually have discrete spatial symmetries, such as spatial inversion and n-fold rotation symmetries. The combinations of spatial and translation symmetries in three dimensions generate 230 symmetry groups in total, which is referred to as *space group*. By definition, every space group has a translation group as its subgroup. Let $\mathcal{G}$ be a space group. A group action of $g \in \mathcal{G}$ on $\mathbb{R}^3$ is defined by

$$g(\boldsymbol{x}) = p_g \boldsymbol{x} + \boldsymbol{t}_g \in \mathbb{R}^3, \tag{2.2}$$

where p_g is an element of the three-dimensional orthogonal group $\mathrm{O}(3)$ and $\boldsymbol{t}_g \in \mathbb{R}^3$ is not always a lattice vector. In the following, we often represent $g \in \mathcal{G}$ by $g = \{p_g|\boldsymbol{t}_g\}$. We define the multiplication of two elements g and g' by

$$p_{gg'} = p_g p_{g'}, \quad \boldsymbol{t}_{gg'} = p_g \boldsymbol{t}_{g'} + \boldsymbol{t}_g. \tag{2.3}$$

It should be noted that $\boldsymbol{t}_g$ is not invariant under a change of origin. Then, there are two types of space groups: *symmorphic* and *nonsymmorphic* space groups. A space group $\mathcal{G}$ is symmorphic (nonsymmorphic) when origin can (cannot) be set in such a way that $\boldsymbol{t}_g$ is a lattice translation for all elements of $\mathcal{G}$.

2.1.3 *Magnetic Space Group*

So far, we have considered only spatial symmetries. However, magnetically disordered materials generally have TRS $\mathcal{T}$, which is an antiunitary symmetry. Furthermore, magnetically ordered materials do not have pure TRS but usually have products of elements in $\mathcal{G}$ and $\mathcal{T}$. Such a symmetry group is called *magnetic space group* and is described by

$$\mathcal{M} = \mathcal{G} + \mathcal{A}, \tag{2.4}$$

where $\mathcal{A}$ is a set of antiunitary symmetries. A group action of $a \in \mathcal{A}$ on $\mathbb{R}^3$ is the same as $g \in \mathcal{G}$, i.e.,

$$a(\boldsymbol{x}) \equiv p_a \boldsymbol{x} + \boldsymbol{t}_a. \tag{2.5}$$

There are 1651 magnetic space groups in three spatial dimensions, which are classified into four types. Note that magnetic space groups contain all ordinary space groups, i.e., $\mathcal{A} = \emptyset$. Such magnetic space groups belong to type I. When $\mathcal{A} = \mathcal{G}\mathcal{T}$, the magnetic space group is said to be type II. If $\mathcal{A} = \mathcal{T} g_0 \mathcal{G}$ for $g_0 \notin \mathcal{G}$, the remaining magnetic space groups are further divided into type III or IV. If g_0 is a pure half-translation, the magnetic space group is type IV; otherwise, it is type III.

2.2 Symmetry and Representations in Solids

In this section, we review how to incorporate symmetries of crystals into electronic band structures. In particular, we see how symmetries can be expressed in tight-binding models and how they provide constraints on electronic band structures.

2.2.1 *Tight-Binding Models and Representations*

In solid-state physics, the tight-binding approximation is usually used to describe electronic structures of free or weakly interacting systems. Suppose that there are N_{orb} degrees of freedom, such as spins and orbitals, in the unit cell, which is denoted by σ. An interaction-free tight-binding Hamiltonian in real space is given by

$$\hat{H} = \sum_{\boldsymbol{R}, \delta\boldsymbol{R}} \sum_{\sigma, \sigma'} \hat{c}^{\dagger}_{\sigma, \boldsymbol{R}+\delta\boldsymbol{R}} [h_{\delta\boldsymbol{R}}]_{\sigma\sigma'} \hat{c}_{\sigma', \boldsymbol{R}}, \tag{2.6}$$

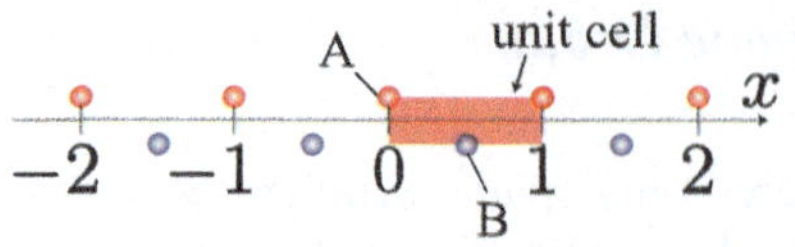

Fig. 2.1 **Real-space illustration of the tight-binding model of** Eq. (2.8). There are two sites at $x = R$ and $x = R + 1/2$ in the unit cell at R, which are denoted by A and B, respectively. The hopping between A and B is $-t$. Also, we add the onsite potential to each site

where $\hat{c}^\dagger_{\sigma,\boldsymbol{R}}$ and $\hat{c}_{\sigma,\boldsymbol{R}}$ represent fermionic creation and annihilation operators at $\boldsymbol{R}$ with the degree of freedom σ such that $\{\hat{c}^\dagger_{\sigma,\boldsymbol{R}}, \hat{c}_{\sigma',\boldsymbol{R}'}\} = \delta_{\sigma,\sigma'}\delta_{\boldsymbol{R},\boldsymbol{R}'}$. Here, we assume the translation symmetries, and thus the matrix element $[h_{\delta\boldsymbol{R}}]_{\sigma\sigma'}$ does not depend on $\boldsymbol{R}$ but only on $\delta\boldsymbol{R}$. Also, we consider that the hopping range $\delta\boldsymbol{R}$ is finite.

Now, we move on to the action of spatial symmetries on electrons. Symmetry $g \in \mathcal{M}$ such that $\hat{g}\hat{H}\hat{g}^{-1} = \hat{H}$ transforms $\hat{c}^\dagger_{\sigma,\boldsymbol{R}}$ by

$$\hat{g}\hat{c}^\dagger_{\sigma,\boldsymbol{R}}\hat{g}^{-1} = \sum_{\sigma'} \hat{c}^\dagger_{\sigma',\tilde{\boldsymbol{R}}_\sigma}[U(g)]_{\sigma'\sigma}, \tag{2.7}$$

where $U(g)$ is a N_{orb}-dimensional unitary matrix representing transformation properties of the degrees of freedom in the unit cell. It should be noted that $\tilde{\boldsymbol{R}}_\sigma \neq g(\boldsymbol{R}) = p_g\boldsymbol{R} + \boldsymbol{t}_g$ in general. For example, let us consider a one-dimensional system with two degrees of freedom in the unit cell (see Fig. 2.1):

$$\hat{H} = \sum_{R=1}^{L}\left[it\left(\hat{c}^\dagger_{B,R}\hat{c}_{A,R} + \hat{c}^\dagger_{A,R+1}\hat{c}_{B,R}\right) + \text{h.c.}\right] + \sum_{R=1}^{L}\left[-\mu\hat{c}^\dagger_{A,R}\hat{c}_{A,R} - \mu\hat{c}^\dagger_{B,R}\hat{c}_{B,R}\right], \tag{2.8}$$

where the periodic boundary condition is assumed. This system is invariant under a screw symmetry $S_x = \left\{C_{2x}|\frac{1}{2}\boldsymbol{e}_x\right\}$ such that

$$S_x : \hat{c}^\dagger_{A,R} \to \hat{c}^\dagger_{B,R} \text{ and } \hat{c}^\dagger_{B,R} \to \hat{c}^\dagger_{A,R+1}, \tag{2.9}$$

where R is the position of a unit cell. While A is transformed to B in the same unit cell, B moves to A in the next unit cell, i.e., $\tilde{R}_A = R$ and $\tilde{R}_B = R + 1$.

2.2.2 *Momentum Space*

In the presence of translation symmetries, it is useful to introduce momentum $\boldsymbol{k}$ such that the eigenvalue of $T_{\boldsymbol{R}} \in T$ is $e^{-i\boldsymbol{k}\cdot\boldsymbol{R}}$. Three-dimensional reciprocal space, called momentum space in the following, is spanned by

$$\boldsymbol{b}_1 = 2\pi\frac{\boldsymbol{a}_2 \times \boldsymbol{a}_3}{\boldsymbol{a}_1 \cdot (\boldsymbol{a}_2 \times \boldsymbol{a}_3)}, \quad \boldsymbol{b}_2 = 2\pi\frac{\boldsymbol{a}_3 \times \boldsymbol{a}_1}{\boldsymbol{a}_2 \cdot (\boldsymbol{a}_3 \times \boldsymbol{a}_1)}, \quad \boldsymbol{b}_3 = 2\pi\frac{\boldsymbol{a}_1 \times \boldsymbol{a}_2}{\boldsymbol{a}_3 \cdot (\boldsymbol{a}_1 \times \boldsymbol{a}_2)}. \tag{2.10}$$

From Eq. (2.10), we find that $e^{-i(\boldsymbol{k}+\boldsymbol{G})\cdot\boldsymbol{R}} = e^{-i\boldsymbol{k}\cdot\boldsymbol{R}}$ for a reciprocal lattice vector $\boldsymbol{G} = \sum_{i=1}^{3} m_i \boldsymbol{b}_i$ ($m_i \in \mathbb{Z}$). Thus, it is enough to consider a fundamental domain of momentum space, known as the first Brillouin zone. In the following, the first Brillouin zone is shortly denoted by BZ.

Fourier transformation of the creation operator is given by

$$\hat{c}^{\dagger}_{\sigma,\boldsymbol{k}} \equiv \frac{1}{\sqrt{V}} \sum_{\boldsymbol{R}} \hat{c}^{\dagger}_{\sigma,\boldsymbol{R}} e^{i\boldsymbol{k}\cdot\boldsymbol{R}}, \tag{2.11}$$

where V is the volume of the system.[1] One can explicitly confirm that

$$T_{\boldsymbol{R}} \hat{c}^{\dagger}_{\sigma,\boldsymbol{k}} T_{\boldsymbol{R}}^{-1} = e^{-i\boldsymbol{k}\cdot\boldsymbol{R}} \hat{c}^{\dagger}_{\sigma,\boldsymbol{k}}. \tag{2.12}$$

Substituting Eq. (2.11) into Eq. (2.6), we find

$$\begin{aligned} \hat{H} &= \sum_{\boldsymbol{k}} \sum_{\sigma,\sigma'} \hat{c}^{\dagger}_{\sigma,\boldsymbol{k}} \left(\sum_{\delta\boldsymbol{R}} [h_{\delta\boldsymbol{R}}]_{\sigma\sigma'} e^{-i\boldsymbol{k}\cdot\delta\boldsymbol{R}} \right) \hat{c}_{\sigma',\boldsymbol{k}}, \\ &\equiv \sum_{\boldsymbol{k}} \sum_{\sigma,\sigma'} \hat{c}^{\dagger}_{\sigma,\boldsymbol{k}} [H_{\boldsymbol{k}}]_{\sigma\sigma'} \hat{c}_{\sigma',\boldsymbol{k}}, \end{aligned} \tag{2.13}$$

where $H_{\boldsymbol{k}}$ is a N_{orb}-dimensional Hermitian matrix referred to as Bloch Hamiltonian.

Next, we discuss actions of spatial operations on electrons. Again, from Eqs. (2.7) and (2.11), we obtain

$$\hat{g} \hat{c}^{\dagger}_{\sigma,\boldsymbol{k}} \hat{g}^{-1} = \sum_{\sigma'} \hat{c}^{\dagger}_{\sigma',\phi_g p_g \boldsymbol{k}} [U_{\boldsymbol{k}}(g)]_{\sigma'\sigma}, \tag{2.14}$$

where $U_{\boldsymbol{k}}(g)$ is a N_{orb}-dimensional unitary matrix.[2] Here, we introduce a map $\phi : G \to \mathbb{Z}_2 = \{\pm 1\}$, which represents g is unitary (antiunitary) for $\phi_g = +1$ ($\phi_g = -1$). For later convenience, we define a symmetry representation of $g \in \mathcal{M}$ by

---

[1] There are two definitions for the Fourier transformation. In this thesis, we always use the definition in Eq. (2.11) for which the Bloch Hamiltonian is periodic under $\boldsymbol{k} \to \boldsymbol{k} + \boldsymbol{G}$.

[2] $[U_{\boldsymbol{k}}(g)]_{\sigma'\sigma}$ is obtained by $[U(g)]_{\sigma'\sigma}$ and $\tilde{\boldsymbol{R}}_\sigma$. When $\tilde{\boldsymbol{R}}_\sigma = p_g \boldsymbol{R} + \boldsymbol{v}_g$,

$$\begin{aligned} \hat{g} \hat{c}^{\dagger}_{\sigma,\boldsymbol{k}} \hat{g}^{-1} &= \sum_{\boldsymbol{R}} \hat{g} \hat{c}^{\dagger}_{\sigma,\boldsymbol{R}} \hat{g}^{-1} e^{i\phi_g \boldsymbol{k}\cdot\boldsymbol{R}} \\ &= \sum_{\sigma'} \sum_{\boldsymbol{R}} \hat{c}^{\dagger}_{\sigma',\tilde{\boldsymbol{R}}_\sigma} [U(g)]_{\sigma'\sigma} e^{i\phi_g \boldsymbol{k}\cdot\boldsymbol{R}} \\ &= \sum_{\sigma'} \sum_{\tilde{\boldsymbol{R}}_\sigma} \hat{c}^{\dagger}_{\sigma',\tilde{\boldsymbol{R}}_\sigma} [U(g)]_{\sigma'\sigma} e^{i\phi_g \boldsymbol{k}\cdot(p_g^{-1}(\tilde{\boldsymbol{R}}_\sigma - \boldsymbol{v}_g))} \\ &= \sum_{\sigma'} \hat{c}^{\dagger}_{\sigma',p_g\boldsymbol{k}} [U(g)]_{\sigma'\sigma} e^{-i\phi_g (p_g\boldsymbol{k})\cdot\boldsymbol{v}_g}. \end{aligned} \tag{2.15}$$

$$\rho_{\boldsymbol{k}}(g) \equiv \begin{cases} U_{\boldsymbol{k}}(g) & \text{for } \phi_g = +1 \\ U_{\boldsymbol{k}}(g)K & \text{for } \phi_g = -1 \end{cases}, \tag{2.16}$$

where K denotes the complex conjugation operator.

Since a symmetry $g \in \mathcal{M}$ should satisfies the relation $\hat{g}\hat{H}\hat{g}^{-1} = \hat{H}$, we find a transformation of $g \in \mathcal{M}$ for the Bloch Hamiltonian

$$\rho_{\boldsymbol{k}}(g)H_{\boldsymbol{k}} = H_{\phi_g p_g \boldsymbol{k}}\rho_{\boldsymbol{k}}(g). \tag{2.17}$$

For example, let us revisit the model defined in Eqs. (2.8) and (2.9). After performing the Fourier transformation, we get

$$H_k = \begin{pmatrix} -\mu & -it(1-e^{-ik}) \\ it(1-e^{ik}) & -\mu \end{pmatrix}, \tag{2.18}$$

$$\rho_k(S_x) = U_k(S_x) = \begin{pmatrix} 0 & e^{-ik} \\ 1 & 0 \end{pmatrix}. \tag{2.19}$$

It should be noted that the matrices are generally projective representations, i.e.,

$$\rho_{p_{g'}\boldsymbol{k}}(g)\rho_{\boldsymbol{k}}(g') = z_{g,g'}\rho_{\boldsymbol{k}}(gg') \quad \text{for} \quad g, g' \in \mathcal{M}, \tag{2.20}$$

where $z_{g,g'} = \pm 1$ is a projective factor that originates from spins. [3] Because of associative law, projective factors satisfy

$$z_{g,hl}z_{h,l} = z_{g,h}z_{gh,l} \ (g, h, l \in \mathcal{G}). \tag{2.22}$$

One of the most simple examples is the case of spin-rotations. For example, π-rotations of spins along x- and y-axis anticommute with each other, which indicates that the representations are projective. On the other hand, if the system has spin-rotation symmetry or is spinless, we can always choose projective factors such that $z_{g,g'} = +1$ for all elements of $\mathcal{M}$.

Some elements of a magnetic space group transform momentum $\boldsymbol{k}$ into other momenta. When $\phi_g p_g \boldsymbol{k} = \boldsymbol{k} + \boldsymbol{G}$, we say $\boldsymbol{k}$ is invariant under $g \in \mathcal{M}$. Collecting all elements of $\mathcal{M}$ that keep $\boldsymbol{k}$ invariant, we define a subgroup of $\mathcal{M}$

$$\mathcal{M}_{\boldsymbol{k}} = \left\{h \in \mathcal{M} \mid h(\boldsymbol{k}) = \phi_h p_h \boldsymbol{k} = \boldsymbol{k} + \boldsymbol{G}\right\}, \tag{2.23}$$

$$= \mathcal{G}_{\boldsymbol{k}} + \mathcal{A}_{\boldsymbol{k}}, \tag{2.24}$$

---

[3] We can explicitly write down this equation as

$$\rho_{\phi_{g'}p_{g'}\boldsymbol{k}}(g)\rho_{\boldsymbol{k}}(g') = z_{g,g'}\rho_{\boldsymbol{k}}(gg') \Leftrightarrow \begin{cases} U_{p_{g'}\boldsymbol{k}}(g)U_{\boldsymbol{k}}(g') = z_{g,g'}U_{\boldsymbol{k}}(gg')K^{\frac{1-\phi_{g'}}{2}} & \text{for } \phi_g = +1 \\ U_{p_{g'}\boldsymbol{k}}(g)[U_{\boldsymbol{k}}(g')]^* = z_{g,g'}U_{\boldsymbol{k}}(gg')K^{\frac{1+\phi_{g'}}{2}} & \text{for } \phi_g = -1 \end{cases}. \tag{2.21}$$

which is referred to as *little group*. Here, $\mathcal{G}_{\boldsymbol{k}}$ and $\mathcal{A}_{\boldsymbol{k}}$ are the unitary and antiunitary part of $\mathcal{M}_{\boldsymbol{k}}$. Since $\boldsymbol{t}_g$ in Eq. (2.2) does not act on momentum space, little groups always contain the translation group T. When $\mathcal{M}_{\boldsymbol{k}}$ is not the translation group T, $\boldsymbol{k}$ is said to be high-symmetry momentum. In particular, we say that $\boldsymbol{k}$ is a high-symmetry point if every point in a neighborhood of $\boldsymbol{k}$ has a lower symmetry group than $\mathcal{M}_{\boldsymbol{k}}$. Similarly, $\boldsymbol{k}$ is a point belonging to a high-symmetry line (plane) if a neighborhood of $\boldsymbol{k}$ contains a line (plane) whose symmetry group is the same as $\mathcal{M}_{\boldsymbol{k}}$ [7].

Since $\mathcal{M}_{\boldsymbol{k}}$ is an infinite group due to its translation subgroup, it is often convenient to introduce a finite group $\mathcal{M}_{\boldsymbol{k}}/T$, which is isomorphic to a magnetic point group. In the literature, $\mathcal{M}_{\boldsymbol{k}}/T$ is referred to as *little co-group* [7]. We can always relate representations of $\mathcal{M}_{\boldsymbol{k}}$ to those of $\mathcal{M}_{\boldsymbol{k}}/T$ by defining the representation $\sigma_{\boldsymbol{k}}(g)$ of $\mathcal{M}_{\boldsymbol{k}}/T$ as

$$\sigma_{\boldsymbol{k}}(g) = \begin{cases} U_{\boldsymbol{k}}(g)e^{-i\boldsymbol{k}\cdot\boldsymbol{t}_g} & \text{for } \phi_g = +1 \\ U_{\boldsymbol{k}}(g)e^{-i\boldsymbol{k}\cdot\boldsymbol{t}_g}K & \text{for } \phi_g = -1 \end{cases}, \tag{2.25}$$

where $\boldsymbol{t}_g$ is a fractional translation or zeros [10]. Correspondingly, projective factors also change as

$$\sigma_{\boldsymbol{k}}(g)\sigma_{\boldsymbol{k}}(h) = z^{\boldsymbol{k}}_{g,h}\sigma_{\boldsymbol{k}}(gh), \tag{2.26}$$

where $z^{\boldsymbol{k}}_{g,h} = z_{g,h}e^{-i\boldsymbol{k}\cdot(p_g\boldsymbol{t}_h-\phi_g\boldsymbol{t}_h)}$. It should be noted that the additional factors originating from nonsymmorphic symmetries lead to nontrivial effects on irreducible representations of little groups. We see an example in Sect. 2.3.2.

2.2.3 Band Structure

Once we have the Bloch Hamiltonian $H_{\boldsymbol{k}}$, what we have to do is to solve the following eigenvalue problem

$$H_{\boldsymbol{k}}\psi_{\boldsymbol{k};ni} = \epsilon_{\boldsymbol{k};n}\psi_{\boldsymbol{k};ni}, \tag{2.27}$$

where $\epsilon_{\boldsymbol{k};n}$ is the nth eigenenergy at $\boldsymbol{k}$ and $\psi_{\boldsymbol{k};ni}$ is an eigenvector of $H_{\boldsymbol{k}}$. Note that, since some eigenvalues of $H_{\boldsymbol{k}}$ are often degenerate, $\psi_{\boldsymbol{k};ni}$ has an index i that represents the ith eigenvector with the eigenvalue $\epsilon_{\boldsymbol{k};n}$. Then, the Bloch Hamiltonian is diagonalized as

$$H_{\boldsymbol{k}} = \Psi_{\boldsymbol{k}}\begin{pmatrix} \epsilon_{\boldsymbol{k};1}\mathbb{1}_{\alpha_1} & 0 & \cdots & 0 \\ 0 & \epsilon_{\boldsymbol{k};2}\mathbb{1}_{\alpha_2} & \cdots & 0 \\ \vdots & \vdots & \ddots & \cdots \\ 0 & 0 & \cdots & \epsilon_{\boldsymbol{k};m_{\boldsymbol{k}}}\mathbb{1}_{\alpha_{m_{\boldsymbol{k}}}} \end{pmatrix}\Psi_{\boldsymbol{k}}^{\dagger}, \tag{2.28}$$

where $\Psi_{\boldsymbol{k}}$ is composed of the set of eigenvectors $\{\psi_{\boldsymbol{k};ni}\}$ and the dimension of $\mathbb{1}_{\alpha_n}$ is the same as degeneracy of $\epsilon_{\boldsymbol{k};n}$. The set of $\{\epsilon_{\boldsymbol{k};m_{\boldsymbol{k}}}\}_{m_{\boldsymbol{k}},\boldsymbol{k}}$ is called *band structure*. Suppose that the system has ν electrons per unit cell. We say that a system is insulating when the system satisfies the following conditions for arbitrary $\boldsymbol{k}$: (a) the νth energy $\epsilon_{\boldsymbol{k};\nu}$ and the $(\nu+1)$th energy $\epsilon_{\boldsymbol{k};(\nu+1)}$ are completely separated and (b) $\epsilon_{\boldsymbol{k};\nu}$ is smaller than the Fermi energy.[4]

As seen in Eq. (2.17), $H_{\boldsymbol{k}}$ does not always commute with $U_{\boldsymbol{k}}(g)$ while $\hat{H}$ does. However, when we consider $h \in \mathcal{G}_{\boldsymbol{k}}$, $H_{\boldsymbol{k}}U_{\boldsymbol{k}}(h) - U_{\boldsymbol{k}}(h)H_{\boldsymbol{k}} = 0$. As discussed in Ref. [9], we can choose a basis such that $H_{\boldsymbol{k}}$ and $U_{\boldsymbol{k}}(h)$ are simultaneously block-diagonalized, i.e.,

$$U_{\boldsymbol{k}}(h) = \mathrm{diag}\left[U_{\boldsymbol{k}}^{\alpha_1}(h) \otimes \mathbb{1}_{N_1}, \dots, U_{\boldsymbol{k}}^{\alpha_n}(h) \otimes \mathbb{1}_{N_n}\right], \tag{2.29}$$

$$H_{\boldsymbol{k}} = \mathrm{diag}\left[\mathbb{1}_{D^{\alpha_1}} \otimes H_{\boldsymbol{k}}^{\alpha_1}, \dots, \mathbb{1}_{D^{\alpha_n}} \otimes H_{\boldsymbol{k}}^{\alpha_n}\right], \tag{2.30}$$

where $U_{\boldsymbol{k}}^{\alpha}(h)$ is an irreducible representations of $\mathcal{G}_{\boldsymbol{k}}$. Here, D^{α} and N_{α} are dimensions of $U_{\boldsymbol{k}}^{\alpha}(h)$ and $H_{\boldsymbol{k}}^{\alpha}$, respectively. Equations (2.29) and (2.30) imply that the Hamiltonian at $\boldsymbol{k}$ can always be decomposed into each irreducible representation sector, which will be used in Chap. 3.

As is the case for the relation between $\rho_{\boldsymbol{k}}(g)$ and $\sigma_{\boldsymbol{k}}(g)$, an irreducible representation $u_{\boldsymbol{k}}^{\alpha}$ of $\mathcal{G}_{\boldsymbol{k}}/T$ is simply related to the irreducible representations $U_{\boldsymbol{k}}^{\alpha}$ of $\mathcal{G}_{\boldsymbol{k}}$ by

$$U_{\boldsymbol{k}}^{\alpha}(g) = u_{\boldsymbol{k}}^{\alpha}(g)e^{-i\boldsymbol{k}\cdot\boldsymbol{t}_g}. \tag{2.31}$$

2.3 Symmetry Relations Among Irreducible Representations

Thus far, we have discussed representations at a single momentum. In the presence of symmetries, irreducible representations are related to each other, which leads to various important consequences. In this section, we review four essential symmetry relations among irreducible representations of little groups: (i) relationship between irreducible representations at two symmetry-related momenta, (ii) degeneracy originating from antiunitary symmetries, (iii) appearance of irreducible representations originating from nonsymmorphic symmetry, and (iv) compatibility relations.

[4] For superconductors, when the system satisfies these conditions, we say the superconductor is gapped.

2.3.1 Relationship Between Irreducible Representations at Two Symmetry-Related k-Points

By definition, $\mathcal{M}_{\boldsymbol{k}}$ is a subgroup of $\mathcal{M}$. Then, $\boldsymbol{k}$ would be transformed by g_l such that $g_l \notin \mathcal{M}_{\boldsymbol{k}}$ but $g_l \in \mathcal{M}$. To see the relation between representations at $\boldsymbol{k}$ and $\boldsymbol{k}_l = \phi_{g_l} p_{g_l} \boldsymbol{k}$, here we discuss $\rho_{\boldsymbol{k}}(g_l)\psi_{\boldsymbol{k};ni}$.

Since $T_{\boldsymbol{R}} g_l = g_l T_{p_{g_l}^{-1}\boldsymbol{R}}$ and $z_{g,T_{\boldsymbol{R}}} = +1$,

$$\begin{aligned}\rho_{g_l\boldsymbol{k}}(T_{\boldsymbol{R}})[\rho_{\boldsymbol{k}}(g_l)\psi_{\boldsymbol{k};ni}] &= \rho_{\boldsymbol{k}}(g_l)\rho_{\boldsymbol{k}}(T_{p_{g_l}^{-1}\boldsymbol{R}})\psi_{\boldsymbol{k};ni} \\ &= \rho_{\boldsymbol{k}}(g_l)\rho_{p_{g_l}\boldsymbol{k}}(T_{\boldsymbol{R}})\psi_{\boldsymbol{k};ni} \\ &= e^{-i(\phi_{g_l} p_{g_l}\boldsymbol{k})\cdot\boldsymbol{R}}[\rho_{\boldsymbol{k}}(g_l)\psi_{\boldsymbol{k};ni}]\end{aligned} \tag{2.32}$$

and

$$\begin{aligned}H_{\phi_{g_l} p_{g_l}\boldsymbol{k}}[\rho_{\boldsymbol{k}}(g_l)\psi_{\boldsymbol{k};ni}] &= \rho_{\boldsymbol{k}}(g_l)H_{\boldsymbol{k}}\psi_{\boldsymbol{k};ni} \\ &= \epsilon_{\boldsymbol{k};n}[\rho_{\boldsymbol{k}}(g_l)\psi_{\boldsymbol{k};ni}].\end{aligned} \tag{2.33}$$

Thus, $\rho_{\boldsymbol{k}}(g_l)\psi_{\boldsymbol{k};ni}$ is a simultaneous eigenstate of $H_{\phi_{g_l} p_{g_l}\boldsymbol{k}}$ and $U_{\phi_{g_l} p_{g_l}\boldsymbol{k}}(T_{\boldsymbol{R}})$ with eigenvalues $\epsilon_{\boldsymbol{k};n}$ and $e^{-i(\phi_{g_l} p_{g_l}\boldsymbol{k})}$. This implies that $\rho_{\boldsymbol{k}}(g_l)\psi_{\boldsymbol{k};ni}$ can be expanded by $\{\psi_{\phi_{g_l} p_{g_l}\boldsymbol{k};ni'}\}_{i'}$ as

$$\rho_{\boldsymbol{k}}(g_l)\psi_{\boldsymbol{k};ni} = \sum_{i'} \psi_{\phi_{g_l} p_{g_l}\boldsymbol{k};ni'}\,[W_{\boldsymbol{k}n}(g_l)]_{i'i}\,, \tag{2.34}$$

where $W_{\boldsymbol{k}n}(g_l)$ is a (dim $U_{\boldsymbol{k}}^{\alpha_n}$)-dimensional unitary matrix, which is called *sewing matrix*. Correspondingly, irreducible representations of little groups at $\boldsymbol{k}$ and $\boldsymbol{k}_l$ are also related to each other. Suppose that we have $h \in \mathcal{G}_{\boldsymbol{k}}$ and $h' = g_l h g_l^{-1} \in \mathcal{G}_{\boldsymbol{k}_l}$. Then, we can construct irreducible representations $U_{\boldsymbol{k}_l}^{\alpha'}$ of $\mathcal{G}_{\boldsymbol{k}_l}$ by

$$U_{\boldsymbol{k}_l}^{\alpha'}(h') = \begin{cases} \frac{z_{h',g_l}}{z_{g_l,h}}[W_{\boldsymbol{k}n}(g_l)U_{\boldsymbol{k}}^{\alpha}(h)W_{\boldsymbol{k}n}^{\dagger}(g_l)] & \text{for } \phi_{g_l} = +1 \\ \frac{z_{h',g_l}}{z_{g_l,h}}[W_{\boldsymbol{k}n}(g_l)[U_{\boldsymbol{k}}^{\alpha}(h)]^* W_{\boldsymbol{k}n}^{\dagger}(g_l)] & \text{for } \phi_{g_l} = -1 \end{cases}. \tag{2.35}$$

In fact, these relations are important for our construction of Atiyah-Hirzebruch spectral sequence introduced in Chap. 4.

2.3.2 Degeneracy Originating from Antiunitary Symmetries

As discussed in Sect. 2.1.3, materials generally have antiunitary symmetries such as time-reversal symmetry $\mathcal{T}$ and the product of a space group symmetry and time-reversal symmetry. Here, we discuss the effects of these antiunitary on irreducible representations of $\mathcal{G}_{\boldsymbol{k}}$.

Following the same discussions in Sect. 2.2.2, we can derive properties of antiunitary symmetries. Similar to the little group $\mathcal{M}_{\boldsymbol{k}}$, we define an antiunitary part $\mathcal{A}_{\boldsymbol{k}}$ of $\mathcal{M}_{\boldsymbol{k}}$ by

$$\mathcal{A}_{\boldsymbol{k}} \equiv \{a \in \mathcal{A} \mid a(\boldsymbol{k}) = \boldsymbol{k} + \boldsymbol{G}\}. \tag{2.36}$$

Antiunitary symmetries often lead to degeneracy of irreducible representations of $\mathcal{G}_{\boldsymbol{k}}$. To see this, let us begin by discussing two simple examples. Let us suppose that $\mathcal{T}$ satisfies $\mathcal{T}^2 = -1$. Let $|\psi\rangle$ be an eigenstate of the Hamiltonian such that $\hat{H}|\psi\rangle = E|\psi\rangle$. If $|\psi\rangle$ is not degenerate, $\hat{\mathcal{T}}|\psi\rangle = e^{i\theta}|\psi\rangle$ $(\theta \in \mathbb{R})$. Then,

$$\hat{\mathcal{T}}[\hat{\mathcal{T}}|\psi\rangle] = e^{-i\theta}[e^{i\theta}|\psi\rangle] = |\psi\rangle. \tag{2.37}$$

However, this contradicts the fact $\mathcal{T}^2 = -1$. Thus, the eigenstates with the eigenvalue E are degenerate.

We further assume that the system has C_{2x} and that $|\psi\rangle$ is also an eigenstate of $\hat{C}_{2x}$, i.e.,

$$\hat{C}_{2x}|\psi\rangle = \zeta|\psi\rangle \quad (\zeta = \pm i). \tag{2.38}$$

We compute the C_{2x}-eigenvalue of $\hat{\mathcal{T}}|\psi\rangle$, and we find

$$\begin{aligned}\hat{C}_{2x}\left(\hat{\mathcal{T}}|\psi\rangle\right) &= \hat{\mathcal{T}}\left(\hat{C}_{2x}|\psi\rangle\right)\\ &= -\zeta\left(\hat{\mathcal{T}}|\psi\rangle\right).\end{aligned} \tag{2.39}$$

Thus, $|\psi\rangle$ and $\hat{\mathcal{T}}|\psi\rangle$ have the same energy, but they are different states.

More generally, we here discuss the degeneracy of irreducible representations of $\mathcal{G}_{\boldsymbol{k}}$ due to antiunitary symmetries. It is convenient to use irreducible representations of $\mathcal{G}_{\boldsymbol{k}}/T$ for our purpose, which is discussed in Sect. 2.2.2. Let $u_{\boldsymbol{k}}^{\alpha}$ be an irreducible representation of $\mathcal{G}_{\boldsymbol{k}}/T$. Then, we introduce the following quantity [1, 7]:

$$W_{\boldsymbol{k}}^{\alpha}(\mathcal{A}) \equiv \frac{1}{|\mathcal{A}_{\boldsymbol{k}}/T|}\sum_{a\in\mathcal{A}_{\boldsymbol{k}}/T} z_{a,a}^{\boldsymbol{k}}\mathrm{tr}[u_{\boldsymbol{k}}^{\alpha}(a^2)] \tag{2.40}$$

$$= \begin{cases} 0 & \text{(i)} \\ -1 & \text{(ii)} \\ 1 & \text{(iii)} \end{cases}. \tag{2.41}$$

This quantity tells us that antiunitary symmetries change the degeneracy of $u_{\boldsymbol{k}}^{\alpha}$ as follows:

Table 2.1 The product table of $\mathcal{A}_\Gamma/T = \{\mathcal{T}, \mathcal{T}C_{2x}, \mathcal{T}C_{2y}, \mathcal{T}C_{2z}\}$

	$\mathcal{T}$	$\mathcal{T}C_{2x}$	$\mathcal{T}C_{2y}$	$\mathcal{T}C_{2z}$
$\mathcal{T}$	e	C_{2x}	C_{2y}	C_{2z}
$\mathcal{T}C_{2x}$	C_{2x}	e	C_{2z}	C_{2y}
$\mathcal{T}C_{2y}$	C_{2y}	C_{2z}	e	C_{2x}
$\mathcal{T}C_{2z}$	C_{2z}	C_{2y}	C_{2x}	e

Table 2.2 The factor system $z^\Gamma_{a,a'}$ for $a, a' \in \mathcal{A}_\Gamma/T = \{\mathcal{T}, \mathcal{T}C_{2x}, \mathcal{T}C_{2y}, \mathcal{T}C_{2z}\}$

$z^\Gamma_{a,a'}$	$\mathcal{T}$	$\mathcal{T}C_{2x}$	$\mathcal{T}C_{2y}$	$\mathcal{T}C_{2z}$
$\mathcal{T}$	-1	-1	-1	-1
$\mathcal{T}C_{2x}$	-1	1	-1	1
$\mathcal{T}C_{2y}$	-1	1	1	-1
$\mathcal{T}C_{2z}$	-1	-1	1	1

(i) When $W^\alpha_k(\mathcal{A}) = 0$, u^α_k forms a pair with a different irreducible representation[5];
(ii) When $W^\alpha_k(\mathcal{A}) = -1$, two u^α_k are paired;
(iii) When $W^\alpha_k(\mathcal{A}) = +1$, the degeneracy does not change.

This diagnosis is called *Wigner criterion*.[6] Indeed, the above two examples are Case (ii) and (i), respectively. Below, we discuss an example of Case (iii). We consider $\Gamma = (0, 0, 0)$ point in magnetic space group $P2221'$ (equivalent to space group $P222$ with time-reversal symmetry) of spinful systems. The little co-group and the antiunitary part are $\mathcal{G}_\Gamma/T = \{e, C_{2x}, C_{2y}, C_{2z}\}$ and $\mathcal{A}_\Gamma/T = \{\mathcal{T}, \mathcal{T}C_{2x}, \mathcal{T}C_{2y}, \mathcal{T}C_{2z}\}$. There exists only one two-dimensional irreducible representation such that

$$u^5_\Gamma(e) = \mathbb{1}, u^5_\Gamma(C_{2x}) = -i\sigma_x, u^5_\Gamma(C_{2y}) = -i\sigma_y, u^5_\Gamma(C_{2z}) = -i\sigma_z, \tag{2.42}$$

where σ_i $(i = x, y, z)$ are Pauli matrices. The product table and factor system are listed in Tables 2.1 and 2.2. We find that

$$W^\alpha_\Gamma(\mathcal{A}) = \frac{1}{4}(-2 + 2 + 2 + 2) = 1. \tag{2.43}$$

Therefore, there are no additional degeneracies.

In the above examples, we have considered only symmorphic space groups. As discussed in Sect. 2.1, the nonsymmorphic symmetries might lead to different degeneracy of irreducible representations on the boundaries of BZ. At the end of this section,

[5] The different representation is given by $U^{\bar{\alpha}}_k = \frac{z_{g,a}}{z_{a,a^{-1}ga}}[U^\alpha_k(a^{-1}ga)]^*$.

[6] Originally, this notion was introduced for magnetic space groups. However, Eq. (2.40) is applicable to any antiunitary symmetry group. We will apply it to antiunitary symmetries generated by particle-hole symmetry.

let us discuss an example of magnetic space group $P2_11'$ whose unitary part is generated by the screw symmetry $S_y = \{C_{2y} \mid \frac{1}{2}\boldsymbol{e}_y\}$. The discussion for $\Gamma = (0, 0, 0)$ in $P2_11'$ is the same as $P21'$; that is, two different irreducible representations $u_\Gamma^\alpha(S_y) = \alpha i$ $(\alpha = \pm 1)$ pair up. On the other hand, the situation at $\mathrm{Z} = (0, \pi, 0)$ is different. Since $z^{\mathrm{Z}}_{\mathcal{T}S_y,\mathcal{T}S_y} = z_{\mathcal{T}S_y,\mathcal{T}S_y} e^{-i(0,\pi,0)\cdot(\frac{1}{2}\boldsymbol{e}_y+\frac{1}{2}\boldsymbol{e}_y)} = +1 \times (-1) = -1$,

$$W_{\mathrm{Z}}^\alpha(\mathcal{A}) = \frac{1}{2}(-1 \times 1 + (-1) \times 1) = -1. \tag{2.44}$$

Thus, $u_{\mathrm{Z}}^+(S_y) = +1$ and $u_{\mathrm{Z}}^-(S_y) = -1$ no longer pair up but form a pair by themselves.

2.3.3 *Appearance of Irreducible Representations Originating from Nonsymmorphic Symmetry*

In this subsection, we see that nonsymmorphic symmetries give rise to constraints on the appearance of irreducible representations [10]. While $H_{\boldsymbol{k}}$ and $U_{\boldsymbol{k}}(g)$ $(g \in \mathcal{G})$ are periodic under $\boldsymbol{k} \to \boldsymbol{k} + \boldsymbol{G}$, irreducible representations are often not in the presence of nonsymmorphic symmetries, i.e., $U_{\boldsymbol{k}+\boldsymbol{G}}^\alpha(g) = U_{\boldsymbol{k}}^\alpha(g)e^{-i\boldsymbol{G}\cdot\boldsymbol{t}_g} \neq U_{\boldsymbol{k}}^\alpha(g)$. Consequently, multiple different irreducible representations appear together, and they intersect with each other at some points as momentum $\boldsymbol{k}$ goes to $\boldsymbol{k} + \boldsymbol{G}$ [10]. Furthermore, we show that these properties provide criteria for the existence of the band gap.

To see how such a relation appears, let us discuss a quasi-one-dimensional spinless system with screw symmetry S_{nx} (n-fold rotation with fractional translation $t_{S_{nx}} = (1/n, 0, 0)^T$) [10]. For example, the system described by Eqs. (2.18) and (2.19) is the case of $n = 2$. Since $\mathcal{G}_k/T$ is isomorphic to a point group generated by n-fold rotation, there are n irreducible representations of $\mathcal{G}_k/T$: $u_k^\alpha(S_{nx}) = e^{i\alpha 2\pi/n}$ $(\alpha = 0, 1, \ldots, n-1)$. From Eq. (2.31), we find irreducible representations of $\mathcal{G}_k$: $U_k^\alpha(S_x) = u_k^\alpha(S_{nx})e^{-ik/n} = e^{i(\alpha 2\pi - k)/n}$. These irreducible representations are not periodic under $k \to k + 2m\pi$ $(m \in \{1, 2, \ldots, n-1\})$ as seen in $U_{k+2m\pi}^\alpha(S_{nx}) = e^{i((\alpha-m)2\pi-k)/n} = U_k^{\alpha-m}(S_{nx})$. This relation implies that a band with $U_k^\alpha(S_{nx})$ must intersect bands with the different irreducible representations between $k = -\pi$ and π.

As mentioned above, these properties of the irreducible representations lead to filling constraints on the existence of the band gap. For simplicity, here we focus on the case of $n = 2$. When the numbers of occupied bands with $U_k^0(S_x)$ and $U_k^1(S_x)$ are different, a band gap closing must occur (see Fig. 2.2). From the above discussion, one can find a constraint on filling; that is, the system must be gapless if the filling is odd in the presence of screw symmetry S_x. Such filling constraints in space groups and magnetic space groups are discussed in Refs. [10–12].

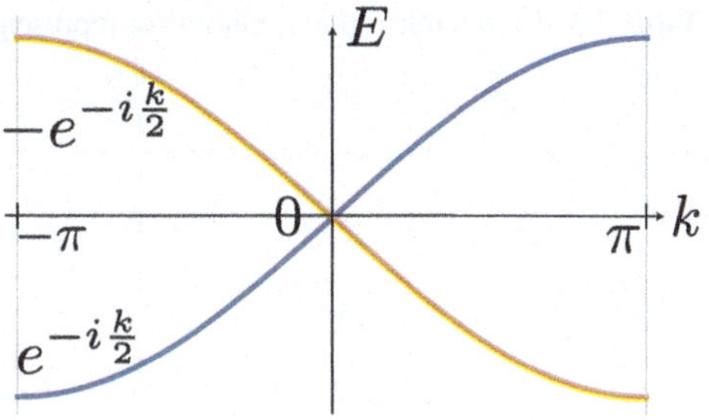

Fig. 2.2 Illustration of the tight-binding model of Eq. (2.8). The band structure of the Bloch Hamiltonian in Eq. (2.18) with $t = 1$ and $\mu = 0$

2.3.4 *Compatibility Relations*

Each energy level should be connected continuously over the entire BZ. This connectivity leads to relationships among irreducible representations of little groups furnished by Bloch wave functions, which are called *compatibility relation* [7]. Importantly, such constraints provide necessary conditions for the existence of a band gap [1, 5, 13], as we will see in the next section. We devote this subsection to a review of compatibility relations that will play a crucial role in symmetry-indicator theory in Sect. 2.4 and the classification of gapless points in Chap. 4.

Before moving on to the general discussion, we start with an example of compatibility relations. Let us consider a quasi one-dimensional system along C_{2x}-rotation axis. The entire BZ is symmetric under C_{2x}, and thus C_{2x}-eigenvalues are defined for each Bloch state at each momentum. Formally, as shown in Eqs. (2.29) and (2.30), the Hamiltonian and the representation can be written as

$$U_k(C_{2x}) = \begin{pmatrix} U_k^{+i}(C_{2x})\mathbb{1}_{\dim H_k^{+i}} & 0 \\ 0 & U_k^{-i}(C_{2x})\mathbb{1}_{\dim H_k^{-i}} \end{pmatrix} = \begin{pmatrix} i\mathbb{1}_{\dim H_k^{+i}} & 0 \\ 0 & (-i)\mathbb{1}_{\dim H_k^{-i}} \end{pmatrix}, \quad (2.45)$$

$$H_k = \begin{pmatrix} H_k^{+i} & 0 \\ 0 & H_k^{-i} \end{pmatrix}. \quad (2.46)$$

Suppose that we have $H_{k=0}^{+i} = -\mathbb{1}_n$ and $H_{k=0}^{-i} = \mathbb{1}_m$ ($n, m \in \mathbb{Z}$). This implies that all occupied bands at $k = 0$ have twofold rotation eigenvalues $+i$. Then, let us consider infinitesimally small momentum δk. Obviously, all occupied bands at $k = \delta k$ should have twofold rotation eigenvalues $+i$ to connect all bands at $k = \delta k$ to those at $k = 0$.

We then generalize the above example. Let $\boldsymbol{k}'$ be momentum slightly off from high-symmetry momentum $\boldsymbol{k}$. By definition, $\mathcal{G}_{\boldsymbol{k}'} \subseteq \mathcal{G}_{\boldsymbol{k}}$. When we restrict $\mathcal{G}_{\boldsymbol{k}}$ to $\mathcal{G}_{\boldsymbol{k}'}$, an irreducible representation of $\mathcal{G}_{\boldsymbol{k}}$ is decomposed into those of $\mathcal{G}_{\boldsymbol{k}'}$, i.e., $U_{\boldsymbol{k}}^{\alpha} \downarrow \mathcal{G}_{\boldsymbol{k}'} = \bigoplus_{\beta} c_{\boldsymbol{k},\boldsymbol{k}'}^{\alpha\beta} U_{\boldsymbol{k}'}^{\beta}$. In the above example, $\mathcal{G}_{k=0} = \mathcal{G}_{k=\delta k}$ and $U_{k=0}^{\pm i} \downarrow \mathcal{G}_{k=\delta k} = U_{k=\delta k}^{\pm i}$.

Let us look into another example of a high-symmetry line $\Lambda = (0, 0, \Lambda)$ connecting $\Gamma = (0, 0, 0)$ to $Z = (0, 0, \pi)$ in space group $P222$. While the little co-groups are $\mathcal{G}_{\Gamma}/T = \mathcal{G}_{Z}/T = \{e, C_{2x}, C_{2y}, C_{2z}\}$, $\mathcal{G}_{\Lambda}/T = \{e, C_{2z}\}$. We first consider spinless electrons; that is, $z_{g,g'} = +1$ for $\forall g, g' \in \mathcal{G}$. Correspondingly, whereas there are

Table 2.3 Irreducible representations for representative elements of $\mathcal{G}_\Gamma$ and $\mathcal{G}_Z$

	e	C_{2x}	C_{2y}	C_{2z}
Γ_1 (Z_1)	1	1	1	1
Γ_2 (Z_2)	1	1	1	1
Γ_3 (Z_3)	1	1	1	-1
Γ_4 (Z_4)	1	1	1	-1
Γ_5 (Z_5)	$\begin{pmatrix} 1 & 0 \\ 0 & 1 \end{pmatrix}$	$\begin{pmatrix} 0 & -i \\ -i & 0 \end{pmatrix}$	$\begin{pmatrix} 0 & -1 \\ 1 & 0 \end{pmatrix}$	$\begin{pmatrix} -i & 0 \\ 0 & i \end{pmatrix}$

Γ_i and Z_i for $i = 1, 2, 3, 4$ are representations for spinless electrons, and Γ_5 and Z_5 are those for spinful electrons. Reproduced from Ref. [14] with some modifications

Table 2.4 Irreducible representations for representative elements of $\mathcal{G}_\Lambda$

	e	C_{2z}
Λ_1	1	1
Λ_2	1	-1
Λ_3	1	i
Λ_4	1	$-i$

Λ_i for $i = 1, 2$ are representations for spinless electrons, and Λ_i for $i = 3, 4$ are those for spinful electrons. Reproduced from Ref. [14]

four one-dimensional irreducible representations at Γ and Z [denoted by Γ_i and Z_i for $i = 1, 2, 3, 4$ in Table 2.3], there are two one-dimensional irreducible representations at Λ [denoted by Λ_j for $j = 1, 2$ in Table 2.4]. When we focus on $\mathcal{G}_\Lambda$-part of $\mathcal{G}_\Gamma$ or $\mathcal{G}_Z$, the two sets of irreducible representations at Γ, (Γ_1, Γ_2) and (Γ_3, Γ_4), cannot be distinguished by C_{2z}-eigenvalues, which implies

$$\Gamma_1 \downarrow \mathcal{G}_\Lambda = \Gamma_2 \downarrow \mathcal{G}_\Lambda = \Lambda_1, \tag{2.47}$$

$$\Gamma_3 \downarrow \mathcal{G}_\Lambda = \Gamma_4 \downarrow \mathcal{G}_\Lambda = \Lambda_2. \tag{2.48}$$

The same relations hold for irreducible representations at Z. Similarly, for spinful electrons, we have

$$\Gamma_5 \downarrow \mathcal{G}_\Lambda = \Lambda_3 \oplus \Lambda_4, \tag{2.49}$$

$$Z_5 \downarrow \mathcal{G}_\Lambda = \Lambda_3 \oplus \Lambda_4. \tag{2.50}$$

2.4 Abelian Group of Band Structures

For a given $H_{\boldsymbol{k}}$ and symmetry representations of a magnetic space group $\mathcal{M}$, we can compute irreducible representations at each momentum. This process can be considered as a map from the $H_{\boldsymbol{k}}$ to a set of irreducible representations. For convenience, we introduce the number of irreducible representation $U_{\boldsymbol{k}}^\alpha$ appearing in occupied bands at $\boldsymbol{k}$, which is denoted by $n_{\boldsymbol{k}}^\alpha$. Then, such numbers form a list

$$\vec{n} = \left(n_{k_1}^{\alpha_1}, n_{k_1}^{\alpha_2}, \ldots, n_{k_2}^{\beta_1}, n_{k_2}^{\beta_2}, \ldots, n_{k_N}^{\gamma_N}\right). \tag{2.51}$$

Practically, it is enough to collect the counts at high-symmetry points for our purpose. This idea was originally introduced in Ref. [13], and was developed in a more systematic way Ref. [1]. In this section, we review such a characterization discussed in Ref. [1], which will be generalized to the case of superconductors in Chap. 3.

The natural question is whether such numbers are all independent or not. Indeed, for the system to be a band insulator, it must satisfy certain constraints. In the following, we refer to such constraints as *compatibility conditions*.

(i) *Gap condition.*—Since we assume the existence of a band gap, the number of occupied bands should be the same at each momentum. In other words, the integers in $\vec{n}$ should satisfy

$$\nu = \sum_{\alpha} n_{\boldsymbol{k}}^{\alpha} \mathrm{Tr}[U_{\boldsymbol{k}}^{\alpha}(e)] \tag{2.52}$$

for each $\boldsymbol{k}$, where ν is the number of electrons per unit cell. This gives rise to the following relations

$$\sum_{\alpha} n_{\boldsymbol{k}_1}^{\alpha} \mathrm{Tr}[U_{\boldsymbol{k}_1}^{\alpha}(e)] = \sum_{\alpha} n_{\boldsymbol{k}_2}^{\alpha} \mathrm{Tr}[U_{\boldsymbol{k}_2}^{\alpha}(e)] \tag{2.53}$$

for arbitrary pairs of high-symmetry points $(\boldsymbol{k}_1, \boldsymbol{k}_2)$ that we take into account.

(ii) *Symmetry related condition.*—Let g_l be an element of $\mathcal{M}$ but not of $\mathcal{M}_{\boldsymbol{k}}$. Then, g_l changes $\boldsymbol{k}$ to another momentum $\boldsymbol{k}_l = \phi_{g_l} p_{g_l} \boldsymbol{k}$. As discussed in Secs. 2.3.1 and 2.3.2, irreducible representation $U_{\boldsymbol{k}_l}^{\alpha'}$ of $\mathcal{G}_{\boldsymbol{k}_l}$ is induced as

$$U_{\boldsymbol{k}_l}^{\alpha'}(h') = \begin{cases} \frac{z_{h',g_l}}{z_{g_l,h}}[W_{\boldsymbol{k}n}(g_l)U_{\boldsymbol{k}}^{\alpha}(h)W_{\boldsymbol{k}n}^{\dagger}(g_l)] & \text{for } \phi_{g_l} = +1 \\ \frac{z_{h',g_l}}{z_{g_l,h}}[W_{\boldsymbol{k}n}(g_l)[U_{\boldsymbol{k}}^{\alpha}(h)]^* W_{\boldsymbol{k}n}^{\dagger}(g_l)] & \text{for } \phi_{g_l} = -1 \end{cases}, \tag{2.54}$$

where $h \in \mathcal{G}_{\boldsymbol{k}}$ and $h' = g_l h g_l^{-1} \in \mathcal{G}_{\boldsymbol{k}_l}$. Therefore,

$$n_{\boldsymbol{k}_l}^{\alpha'} = n_{\boldsymbol{k}}^{\alpha}. \tag{2.55}$$

This condition will also be used for the case of superconductors as discussed in Chaps. 3 and 4.

(iii) *Band connectivity.*— Let $\boldsymbol{k}'$ be momentum slightly off from high-symmetry momentum $\boldsymbol{k}$. As discussed in Sect. 2.3.4, an irreducible representation of $\mathcal{G}_{\boldsymbol{k}}$ is decomposed into those of $\mathcal{G}_{\boldsymbol{k}'}$, i.e., $U_{\boldsymbol{k}}^{\alpha} \downarrow \mathcal{G}_{\boldsymbol{k}'} = \bigoplus_{\beta} c_{\boldsymbol{k},\boldsymbol{k}'}^{\alpha\beta} U_{\boldsymbol{k}'}^{\beta}$. This implies

$$n_{\boldsymbol{k}'}^{\beta} = \sum_{\alpha} n_{\boldsymbol{k}}^{\alpha} c_{\boldsymbol{k},\boldsymbol{k}'}^{\alpha\beta}, \tag{2.56}$$

where the matrix c can be computed from $\sum_{g\in\mathcal{G}_{k'}}\left(\chi_{k'}^{\beta}(g)\right)^{*}\chi_{k}^{\alpha}(g)$.[7]

All these compatibility conditions can be encoded in an integer matrix C as[8]

$$C\vec{n}=\mathbf{0}. \tag{2.60}$$

Finally, we define an abelian group of band structures that satisfies all compatibility conditions by

$$\{\mathrm{BS}\}\equiv\ker C\cap\mathbb{Z}^{D}\simeq\mathbb{Z}^{d_{\mathrm{BS}}}, \tag{2.61}$$

where $d_{\mathrm{BS}}=\dim\ker C$ [1].

To see how compatibility conditions work, let us consider one- and two-dimensional inversion symmetric systems. Compatibility conditions in this symmetry setting are

$$n_{k_1}^{+}+n_{k_1}^{-}=n_{k_2}^{+}+n_{k_2}^{-}, \tag{2.62}$$

where $(\boldsymbol{k}_1,\boldsymbol{k}_2)=(k=0,k=\pi)$ for 1D and $(\boldsymbol{k}_1,\boldsymbol{k}_2)=(\Gamma,X),(X,M),(M,Y),(Y,\Gamma)$ for 2D. Then, compatibility conditions are expressed by

$$\begin{pmatrix}1 & 1 & -1 & -1\end{pmatrix}\begin{pmatrix}n_{k=0}^{+}\\ n_{k=0}^{-}\\ n_{k=\pi}^{+}\\ n_{k=\pi}^{-}\end{pmatrix}=\vec{0} \tag{2.63}$$

7 A representation $U_{\boldsymbol{k}}(g)$ $(g\in\mathcal{G}_{\boldsymbol{k}'})$ is always expressed as

$$U_{\boldsymbol{k}}(g)=\bigoplus_{\alpha}n_{\boldsymbol{k}}^{\alpha}U_{\boldsymbol{k}}^{\alpha}(g) \tag{2.57}$$

$$=\bigoplus_{\alpha}\bigoplus_{\beta}n_{\boldsymbol{k}}^{\alpha}c_{\boldsymbol{k},\boldsymbol{k}'}^{\alpha\beta}U_{\boldsymbol{k}'}^{\beta}(g). \tag{2.58}$$

On the other hand, since $|\boldsymbol{k}'-\boldsymbol{k}|$ is infinitesimally small,

$$U_{\boldsymbol{k}}(g)=\bigoplus_{\alpha}n_{\boldsymbol{k}'}^{\beta}U_{\boldsymbol{k}'}^{\beta}(g). \tag{2.59}$$

Comparing these two results, we find that $n_{\boldsymbol{k}'}^{\beta}=\sum_{\alpha}n_{\boldsymbol{k}}^{\alpha}c_{\boldsymbol{k},\boldsymbol{k}'}^{\alpha\beta}$.

8 In the previous section, we see that the number of occupied bands with $U_{k}^{\alpha}(S_{nx})$ and $U_{k}^{\alpha-m}(S_{nx})$ must be the same when the system is gapped. This is represented by $n_{k}^{\alpha}=n_{k}^{\alpha-m}$, which should be also considered. Indeed, if we take sufficient numbers of connecting paths into account, such relations originating from nonsymmorphic symmetries are already encoded in the matrix C. Similarly, the first condition on the number of occupied bands is also included in this condition. However, here we explicitly mention it in the main text since we use the first condition in examples.

for 1D and

$$\begin{pmatrix} 1 & 1 & -1 & -1 & 0 & 0 & 0 & 0 \\ 1 & 1 & 0 & 0 & -1 & -1 & 0 & 0 \\ 1 & 1 & 0 & 0 & 0 & 0 & -1 & -1 \end{pmatrix} \begin{pmatrix} n_\Gamma^+ \\ n_\Gamma^- \\ n_X^+ \\ n_X^- \\ n_Y^+ \\ n_Y^- \\ n_M^+ \\ n_M^- \end{pmatrix} = \vec{0} \tag{2.64}$$

for 2D. As a result, we get the following bases of {BS}:

$$\{\mathrm{BS}\} = \mathrm{span}\left\{ \begin{array}{l} \vec{b}_1 = (1\ 0\ 1\ 0)^T \\ \vec{b}_2 = (0\ 1\ 0\ 1)^T \\ \vec{b}_3 = (1\ 0\ 0\ 1)^T \end{array} \right\}. \tag{2.65}$$

and

$$\{\mathrm{BS}\} = \mathrm{span}\left\{ \begin{array}{l} \vec{b}_1 = (1\ 0\ 1\ 0\ 1\ 0\ 1\ 0)^T \\ \vec{b}_2 = (1\ 0\ 1\ 0\ 0\ 1\ 0\ 1)^T \\ \vec{b}_3 = (1\ 0\ 0\ 1\ 1\ 0\ 0\ 1)^T \\ \vec{b}_4 = (0\ 1\ 1\ 0\ 1\ 0\ 0\ 1)^T \\ \vec{b}_5 = (1\ 0\ 1\ 0\ 1\ 0\ 0\ 1)^T \end{array} \right\}. \tag{2.66}$$

for 1D and 2D, respectively.

2.5 Atomic Insulators

Let us assume that all electrons are located at sites without any hopping. We refer to such states as *atomic insulators*. Since these states, by definition, are insulating, their band structures and irreducible representations of little groups are obtained by Fourier transformation. In this section, we review how to map from such a real-space picture to representations in momentum space. In fact, the representations of atomic insulators play a crucial role in the diagnosis of topological insulating phases, as shown in Refs. [1, 5]. Furthermore, they are used for the computations of the symmetry-indicator theory of superconductors, as seen in Chap. 3.

2.5.1 *Representations of Atomic Insulators*

To show procedures for obtaining representations of atomic insulators, let us start with a concrete example of one-dimensional inversion symmetric systems. There

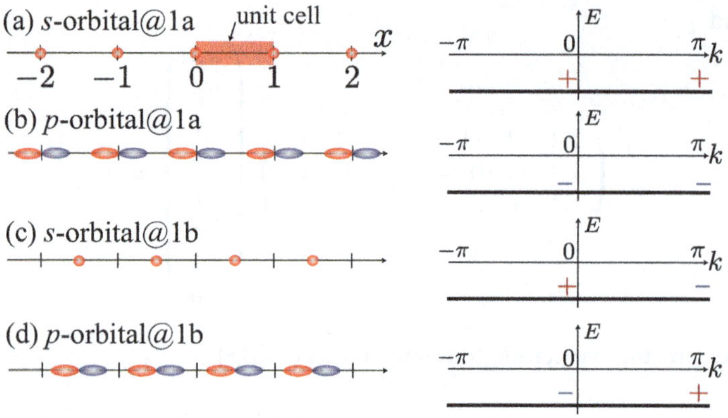

Fig. 2.3 Illustration of 1D atomic insulators in the presence of inversion symmetry. There are two inversion centers, $x = 0$ and $1/2$, in the unit cell and two orbitals, s-orbital (even-parity) and p-orbital (odd-parity). (a) an atomic insulator with the s-orbital at $x = R$ ($R = 0, \pm 1, \pm 2, \ldots$) and its band representations. (b) an atomic insulator with the p-orbital at $x = R$ and its band representations. (c) an atomic insulator with the s-orbital at $x = R + 1/2$ and its band representations. (d) an atomic insulator with the p-orbital at $x = R + 1/2$ and its band representations

are two inversion centers, $x = R$ and $x = R + 1/2$, in the unit cell at $x = R$ [see Fig. 2.3]. As shown in Fig. 2.3a, b, we first put an s- or p-orbital at the inversion center $x = R$. Then, an electron at $x = R$ is transferred as

$$\hat{I}\hat{c}_R^\dagger \hat{I}^{-1} = \xi \hat{c}_{-R}^\dagger, \tag{2.67}$$

where $\xi = +1$ for the s-orbital and $\xi = -1$ for the p-orbital. In momentum space, the electron is transformed as

$$\hat{I}\hat{c}_k^\dagger \hat{I}^{-1} = \frac{1}{\sqrt{V}} \sum_R \hat{I}\hat{c}_R^\dagger \hat{I}^{-1} e^{ikR} \tag{2.68}$$

$$= \xi \frac{1}{\sqrt{V}} \sum_R \hat{c}_{-R}^\dagger e^{i(-k)(-R)} = \xi \hat{c}_{-k}^\dagger \tag{2.69}$$

From this equation, we find $U_k(I) = \xi$.

Next, we put an s- or p-orbital at $x = R + 1/2$. Similar to Eqs. (2.67) and (2.68), we see that the electrons in real and momentum space are transformed as

$$\hat{I}\hat{c}_R^\dagger \hat{I}^{-1} = \xi \hat{c}_{-R-1}^\dagger. \tag{2.70}$$

and

$$\hat{I}\hat{c}_k^\dagger \hat{I}^{-1} = \frac{1}{\sqrt{V}} \sum_R \hat{I}\hat{c}_R^\dagger \hat{I}^{-1} e^{ikR} \tag{2.71}$$

$$= \xi \frac{1}{\sqrt{V}} \sum_R \hat{c}_{-R-1}^\dagger e^{i(-k)(-R-1)} e^{-ik} = \xi \hat{c}_{-k}^\dagger e^{-ik} \tag{2.72}$$

Thus, we have $U_k(I) = \xi e^{-ik}$. It should be noted that the eigenvalues at $k = 0$ and $k = \pi$ are different in this case. Namely, while $U_{k=0}(I) = \xi$, $U_{k=\pi}(I) = -\xi$.

We now move on to discussions about general cases. Here, we follow the discussions of Ref. [3]. Although we characterize electrons by s- and p-orbitals in the above example, they are generally ill-defined. Instead, orbitals at a site $\boldsymbol{r}$ in the unit cell are defined by irreducible representations at $\boldsymbol{r}$.

We first define a site symmetry group of $\boldsymbol{r}$ by

$$\mathcal{G}_{\boldsymbol{r}} \equiv \{h \in \mathcal{G} \mid h(\boldsymbol{r}) = \boldsymbol{r}\}, \tag{2.73}$$

which is a real-space counterpart of the little group $\mathcal{G}_{\boldsymbol{k}}$. Then, the orbitals at $\boldsymbol{r}$ are defined by irreducible representations of $\mathcal{G}_{\boldsymbol{r}}$. In other words, an electron at $\boldsymbol{r}$ is transformed as

$$\hat{h}\hat{c}_{\sigma,\boldsymbol{R}=\boldsymbol{0}}^\dagger \hat{h}^{-1} = \sum_{\sigma'} \hat{c}_{\sigma',\boldsymbol{R}=\boldsymbol{0}}^\dagger [u_{\boldsymbol{r}}^\alpha(h)]_{\sigma'\sigma} \ (h \in \mathcal{G}_{\boldsymbol{r}}), \tag{2.74}$$

where $u_{\boldsymbol{r}}^\alpha$ denotes an irreducible representation of $\mathcal{G}_{\boldsymbol{r}}$.

By definition, $g \in \mathcal{G}\backslash\mathcal{G}_{\boldsymbol{r}}$ will move $\boldsymbol{r}$ to another site. In particular, we collect all sites in the unit cell and label them with $\{\boldsymbol{r}_i\}_i$. Also, we have elements $\{g_i\}_i$ of $\mathcal{G}$ such that $g_1 = e$ and $g_i(\boldsymbol{r}) = \boldsymbol{r}_i$. Then, we define a creation operator at $\boldsymbol{r}_i$ in the unit cell at $\boldsymbol{R}$ by

$$[\hat{c}_{\sigma,\boldsymbol{R}}^\dagger]_i \equiv \hat{T}_{\boldsymbol{R}} \hat{g}_i \hat{c}_{\sigma,\boldsymbol{R}=\boldsymbol{0}}^\dagger \hat{g}_i^{-1} \hat{T}_{\boldsymbol{R}}^{-1}, \tag{2.75}$$

where we explicitly split indices about degrees of freedom into site index i and orbital index σ. Our next task is to see how $[\hat{c}_{\sigma,\boldsymbol{R}}^\dagger]_i$ is transformed by $g \in \mathcal{G}$. For an element $g \in \mathcal{G}$, we can always find $h \in \mathcal{G}_{\boldsymbol{r}}$, $T_{\boldsymbol{R}} \in T$, and $g_i \in$ such that $g = T_{\boldsymbol{R}} g_i h$; that is, $\mathcal{G}$ can be decomposed into left cosets. Since $g\hat{T}_{\boldsymbol{R}} g_i \in \mathcal{G}$, there exist $h \in \mathcal{G}_{\boldsymbol{r}}$, $T_{\boldsymbol{R}} \in T$, and $g_{\bar{i}} \in \mathcal{G}$ such that $g\hat{T}_{\boldsymbol{R}} g_i = T_{\hat{\boldsymbol{R}}} g_{\bar{i}} h$. Taking into account factor systems, we obtain

$$\begin{aligned}\hat{g}[\hat{c}^{\dagger}_{\sigma,\boldsymbol{R}}]_i\hat{g}^{-1} &= \hat{g}\left(\hat{T}_{\boldsymbol{R}}\hat{g}_i\hat{c}^{\dagger}_{\sigma,\boldsymbol{0}}\hat{g}_i^{-1}\hat{T}_{\boldsymbol{R}}^{-1}\right)\hat{g}^{-1}\\ &= z_{g,g_i}\widehat{gT_{\boldsymbol{R}}g_i}\hat{c}^{\dagger}_{\sigma,\boldsymbol{0}}\widehat{gT_{\boldsymbol{R}}g_i}^{-1}\\ &= z_{g,g_i}\widehat{T_{\tilde{\boldsymbol{R}}}g_{\tilde{i}}h}\hat{c}^{\dagger}_{\sigma,\boldsymbol{0}}\widehat{T_{\tilde{\boldsymbol{R}}}g_{\tilde{i}}h}^{-1}\\ &= \frac{z_{g,g_i}}{z_{g_{\tilde{i}},h}}\hat{T}_{\tilde{\boldsymbol{R}}}\hat{g}_{\tilde{i}}\hat{h}\hat{c}^{\dagger}_{\sigma,\boldsymbol{0}}\hat{h}^{-1}\hat{g}_{\tilde{i}}^{-1}\hat{T}_{\tilde{\boldsymbol{R}}}^{-1}\\ &= \sum_{\sigma'}\frac{z_{g,g_i}}{z_{g_{\tilde{i}},h}}[\hat{c}^{\dagger}_{\sigma',\tilde{\boldsymbol{R}}}]_{\tilde{i}}[u^{\alpha}_{\boldsymbol{r}}(h)]_{\sigma'\sigma}.\end{aligned} \tag{2.76}$$

In the same way as the above examples, we perform Fourier transformation and get

$$\hat{g}[\hat{c}^{\dagger}_{\sigma,\boldsymbol{k}}]_i\hat{g}^{-1} = \sum_{\sigma',i'}[\hat{c}^{\dagger}_{\sigma',p_g\boldsymbol{k}}]_{i'}[U^{(\boldsymbol{r},\alpha)}_{\boldsymbol{k}}(h)]_{\sigma'\sigma;i'i} \tag{2.77}$$

$$[U^{(\boldsymbol{r},\alpha)}_{\boldsymbol{k}}(h)]_{\sigma'\sigma;i'i} = \delta'_{\boldsymbol{r}_{i'},g(\boldsymbol{r}_i)}e^{-ip_g\boldsymbol{k}\cdot(g(\boldsymbol{r}_i)-\boldsymbol{r}_{i'})} \times \frac{z_{g,g_i}}{z_{g_{\tilde{i}},h}}[u^{\alpha}_{\boldsymbol{r}}(h)]_{\sigma'\sigma} \tag{2.78}$$

where $h = g_{i'}^{-1}T_{\boldsymbol{r}_{i'}-g(\boldsymbol{r}_i)}gg_i$. Also, $\delta'_{\boldsymbol{x},\boldsymbol{y}} = 1$ only when $\boldsymbol{x}-\boldsymbol{y}$ is a lattice vector; otherwise, $\delta'_{\boldsymbol{x},\boldsymbol{y}} = 0$. One can explicitly confirm that Eq. (2.76) reproduces the results of the above examples.

2.5.2 *Abelian Group of Atomic Insulators*

Once we get $\{U^{(\boldsymbol{r},\beta)}_{\boldsymbol{k}}\}_{\boldsymbol{k}}$ for $(\boldsymbol{r},\beta)$, we can construct $\vec{n}$ in Eq. (2.51) for an atomic insulator

$$\vec{a}_{(\boldsymbol{r},\beta)} = \left(n^{\alpha_1,(\boldsymbol{r},\beta)}_{\boldsymbol{k}_1}, n^{\alpha_2,(\boldsymbol{r},\beta)}_{\boldsymbol{k}_1},\dots,n^{\beta_1,(\boldsymbol{r},\beta)}_{\boldsymbol{k}_2},n^{\beta_2,(\boldsymbol{r},\beta)}_{\boldsymbol{k}_2},\dots,n^{\gamma_N,(\boldsymbol{r},\beta)}_{\boldsymbol{k}_N}\right), \tag{2.79}$$

$$n^{\alpha,(\boldsymbol{r},\beta)}_{\boldsymbol{k}} = \sum_{h\in\mathcal{G}_{\boldsymbol{k}}/T}[U^{\alpha}_{\boldsymbol{k}}(h)]^{*}U^{(\boldsymbol{r},\beta)}_{\boldsymbol{k}}(h) \in \mathbb{Z}_{\geq 0}. \tag{2.80}$$

By collecting $\vec{a}_{(\boldsymbol{r},\beta)}$ for all independent sets of $\{(\boldsymbol{r},\beta)\}_{\beta,\boldsymbol{r}}$, we have the abelian group of atomic insulators

$$\{\mathrm{AI}\} \simeq \mathbb{Z}^{d_{\mathrm{AI}}}, \tag{2.81}$$

which is introduced in Ref. [1]. Practically, it is enough to consider all inequivalent Wyckoff positions that are high-symmetry points in real space.[9]

---

[9] Here, we provide a formal definition of *Wyckoff position*. Let $\mathcal{M}_{\boldsymbol{r}}$ and $\Lambda_{\boldsymbol{r}}$ be a little group of a magnetic space group $\mathcal{M}$ at $\boldsymbol{r}$ and a $\mathcal{M}$-symmetric lattice. Points in real space are classified based on their little group. Two positions $\boldsymbol{r}_1$ and $\boldsymbol{r}_2$ are equivalent if and only if there exists $g \in \mathcal{M}$ such that $\mathcal{M}_{\boldsymbol{r}_2} = g\mathcal{M}_{\boldsymbol{r}_1}g^{-1}$, denoted by $\boldsymbol{r}_1 \sim \boldsymbol{r}_2$. The classification of real space based on this equivalence relation is called the Wyckoff position, which is available in the literature [15].

2.5.3 Examples

Let us start with the one-dimensional inversion symmetric case. As discussed above, we have

$$U_k^{(x=0,\xi)} = \xi, \tag{2.82}$$
$$U_k^{(x=1/2,\xi)} = e^{ik}\xi, \tag{2.83}$$

where $k = 0$ or π and $\xi = \pm 1$ is the orbital at $x = 0$ or $1/2$. From the representations, we have

$$\vec{a}_{(x=0,+)} = (1, 0, 1, 0)^T \tag{2.84}$$
$$\vec{a}_{(x=0,-)} = (0, 1, 0, 1)^T \tag{2.85}$$
$$\vec{a}_{(x=\frac{1}{2},+)} = (1, 0, 0, 1)^T \tag{2.86}$$
$$\vec{a}_{(x=\frac{1}{2},-)} = (0, 1, 1, 0)^T \tag{2.87}$$

There exists a relation $\vec{a}_{(x=0,+)} + \vec{a}_{(x=0,-)} = \vec{a}_{(x=\frac{1}{2},+)} + \vec{a}_{(x=\frac{1}{2},-)}$.[10] This implies that the three of them are independent, and thus, we obtain

$$\{\mathrm{AI}\} = \mathrm{span}\left\{\begin{array}{l} \vec{a}_1 = (1\ 0\ 1\ 0) \\ \vec{a}_2 = (0\ 1\ 0\ 1) \\ \vec{a}_3 = (1\ 0\ 0\ 1) \end{array}\right\} \simeq \mathbb{Z}^3. \tag{2.88}$$

It should be noted that this is equivalent to {BS}.[11]

Next, we explain the two-dimensional inversion symmetric case. Wyckoff positions are $(0, 0)$, $(\frac{1}{2}, 0)$, $(0, \frac{1}{2})$, and $(\frac{1}{2}, \frac{1}{2})$, which are labeled as $1a$, $1b$, $1c$, and $1d$. Site symmetries depend on Wyckoff positions. For example, the site symmetry group of $1a$ is $\{e, I\}$. On the other hand, that of $1b$ is $\{e, T_{\vec{a}_1} I\}$.

From similar discussions about the above one-dimensional case or the general formula in Eq. (2.78), we find

$$U_{\boldsymbol{k}}^{(1a,\xi)} = \xi, \tag{2.89}$$
$$U_{\boldsymbol{k}}^{(1b,\xi)} = e^{ik_x}\xi, \tag{2.90}$$
$$U_{\boldsymbol{k}}^{(1c,\xi)} = e^{ik_y}\xi, \tag{2.91}$$
$$U_{\boldsymbol{k}}^{(1d,\xi)} = e^{ik_x}e^{ik_y}\xi, \tag{2.92}$$

[10] This relation originates from an equivalence relation in real space, which is introduced in Ref. [16]. That is, when s- and p-orbitals sit on $x = 0$, these two orbitals can be moved to $x = 1/2$ in an inversion symmetric manner.

[11] This result will be used in Chap. 3 when we discuss trivial superconductors in one-dimensional inversion symmetric systems.

where $\boldsymbol{k} = \Gamma$, X, Y, or M. Also, $\xi = \pm 1$ corresponds to the s- or p-orbital at the Wyckoff positions. As a result, we have

$$\vec{a}_{(1a,+)} = (1, 0, 1, 0, 1, 0, 1, 0)^T , \tag{2.93}$$

$$\vec{a}_{(1a,-)} = (0, 1, 0, 1, 0, 1, 0, 1)^T , \tag{2.94}$$

$$\vec{a}_{(1b,+)} = (1, 0, 0, 1, 1, 0, 0, 1)^T , \tag{2.95}$$

$$\vec{a}_{(1b,-)} = (0, 1, 1, 0, 0, 1, 1, 0)^T , \tag{2.96}$$

$$\vec{a}_{(1c,+)} = (1, 0, 1, 0, 0, 1, 0, 1)^T , \tag{2.97}$$

$$\vec{a}_{(1c,-)} = (0, 1, 0, 1, 1, 0, 1, 0)^T , \tag{2.98}$$

$$\vec{a}_{(1d,+)} = (1, 0, 0, 1, 0, 1, 1, 0)^T , \tag{2.99}$$

$$\vec{a}_{(1d,-)} = (0, 1, 1, 0, 1, 0, 0, 1)^T . \tag{2.100}$$

Again, they are not independent; that is, $\vec{a}_{(1a,+)} + \vec{a}_{(1a,-)} = \vec{a}_{(1b,+)} + \vec{a}_{(1b,-)} = \vec{a}_{(1c,+)} + \vec{a}_{(1c,-)} = \vec{a}_{(1d,+)} + \vec{a}_{(1d,-)}$. Then, an abelian group of atomic insulators is spanned by

$$\{\mathrm{AI}\} = \mathrm{span}\begin{Bmatrix} \vec{a}_1 = (1\ 0\ 1\ 0\ 1\ 0\ 1\ 0)^T \\ \vec{a}_2 = (0\ 1\ 0\ 1\ 0\ 1\ 0\ 1)^T \\ \vec{a}_3 = (1\ 0\ 0\ 1\ 1\ 0\ 0\ 1)^T \\ \vec{a}_4 = (1\ 0\ 1\ 0\ 0\ 1\ 0\ 1)^T \\ \vec{a}_5 = (1\ 0\ 0\ 1\ 0\ 1\ 1\ 0)^T \end{Bmatrix} \simeq \mathbb{Z}^5. \tag{2.101}$$

We also find that $d_{\mathrm{AI}} = d_{\mathrm{BS}} = 5$.

2.6 Symmetry-Based Indicators for Normal Conducting Phases

In principle, topological phases are identified by topological invariants. Unfortunately, however, we usually do not have complete sets of topological invariants to distinguish all possible phases for a given symmetry setting. Even when we have concrete expressions of topological invariants, it is difficult to compute them by following definitions. Reference [1] has proposed a practical diagnostic scheme that is known as *symmetry-indicator theory*. In this section, we review how to perform it.

In the preceding two sections, we construct two sets of abelian groups of band structures {BS} and {AI}. {BS} includes all topologically nontrivial and trivial band structures that satisfy all compatibility conditions. {AI} is the set of atomic insulators constructed by the method discussed in Sect. 2.5. As a result, $\{\mathrm{AI}\} \subseteq \{\mathrm{BS}\}$, i.e, $d_{\mathrm{AI}} \leq d_{\mathrm{BS}}$. Also, once we have {BS} and {AI}, we can evaluate the mismatch between {BS} and {AI}. Furthermore, after the authors of Ref. [1] computed {BS} and {AI} for all space groups individually, they found that the relation $d_{\mathrm{BS}} = d_{\mathrm{AI}}$ always holds. Therefore, the mismatch is always a finite abelian group and is denoted by

$$X_{\mathrm{BS}} \equiv \{\mathrm{BS}\}/\{\mathrm{AI}\} \simeq \mathbb{Z}_{n_1} \times \cdots \times \mathbb{Z}_{n_N}, \tag{2.102}$$

where we refer to X_{BS} as *symmetry indicator group*. In Ref. [1, 17], X_{BS} for all space groups and magnetic space groups are listed. It should be noted that $d_{\mathrm{BS}} = d_{\mathrm{AI}}$ implies that any $\vec{b} \in \{\mathrm{BS}\}$ can be expanded by a basis set of $\{\mathrm{AI}\}$. However, the coefficients are not always integers but rational numbers. In fact, non-integer coefficients indicate that $\vec{b}$ is not an element of $\{\mathrm{AI}\}$, i.e., the band structure must be topologically nontrivial.

To show how to compute the symmetry indicator group, let us discuss one- and two-dimensional inversion symmetric systems, whose $\{\mathrm{BS}\}$ and $\{\mathrm{AI}\}$ are shown in Sects. 2.4 and 2.5. Let us begin by explaining the one-dimensional case. From Eqs. (2.65) and (2.88), we find

$$(\vec{b}_1, \vec{b}_2, \vec{b}_3) = (\vec{a}_1, \vec{a}_2, \vec{a}_3) \begin{pmatrix} 1 & 0 & 0 \\ 0 & 1 & 0 \\ 0 & 0 & 1 \end{pmatrix}. \tag{2.103}$$

This equation implies that all bases $\{\vec{b}_i\}_{i=1}^{3}$ can be expanded by a basis set of $\{\mathrm{AI}\}$ with integer coefficients. Therefore, $\{\mathrm{BS}\}$ is equivalent to $\{\mathrm{AI}\}$; that is, $X_{\mathrm{BS}} \equiv \{\mathrm{BS}\}/\{\mathrm{AI}\} \simeq 0$. Indeed, Ref. [1] showed that X_{BS} is always trivial in one dimension.

On the other hand, for the two-dimensional case,

$$(\vec{b}_1, \vec{b}_2, \vec{b}_3, \vec{b}_4, \vec{b}_5) = (\vec{a}_1, \vec{a}_2, \vec{a}_3, \vec{a}_4, \vec{a}_5) \begin{pmatrix} 1 & 0 & 0 & 1 & \frac{1}{2} \\ 0 & 0 & 0 & 1 & 0 \\ 0 & 0 & 1 & 0 & \frac{1}{2} \\ 0 & 1 & 0 & 0 & \frac{1}{2} \\ 0 & 0 & 0 & -1 & -\frac{1}{2} \end{pmatrix}. \tag{2.104}$$

From the above equation, $\{\vec{b}_i\}_{i=1}^{4}$ are elements of $\{\mathrm{AI}\}$, but $\vec{b}_5$ cannot be expanded by the basis set of $\{\mathrm{AI}\}$.[12] Furthermore, the expansion coefficients of $2\vec{b}_5$ are all integers. Therefore, we conclude that $X_{\mathrm{BS}} = \mathbb{Z}_2$.

In practice, it is enough to find $\{\mathrm{AI}\}$ for the purpose of computing X_{BS} of normal conducting phases. We can always find a new basis set of $\{\mathrm{AI}\}$ such that

$$(\vec{a}_1, \vec{a}_2, \ldots, \vec{a}_{d_{\mathrm{AI}}})V = U \begin{pmatrix} \Sigma & 0 \\ 0 & 0 \end{pmatrix}, \tag{2.105}$$

where $U = (\vec{u}_1, \vec{u}_2, \ldots)$ and $V = (\vec{v}_1, \vec{v}_2, \ldots)$ are unimodular matrices, and $\Sigma = \mathrm{diag}(s_1, s_2, \ldots s_{d_{\mathrm{AI}}})$ is called *Smith normal form* [4]. Also, s_i's are all positive integers.

[12] While $\vec{b}_4$ is the same representation contents of $\vec{a}_{1d,-}$, sometimes one cannot find any combination of $\{\vec{a}_{(r,\alpha)}\}_{(r,\alpha)}$ such that the $\vec{b}$ of a topologically trivial system is expanded with *non-negative* integer-valued coefficients. This indicates that the system cannot be continuously deformed to any atomic insulator. However, such band topology is fragile against adding trivial degrees of freedom, which is known as *fragile topology* [18, 19]. Interestingly, fragile topology plays an important role in studies on twisted bilayer graphene [20, 21].

Since V is a unimodular matrix, the left-hand side is a new basis set of {AI}. Thus, $s_i u_i$ ($i \leq d_{\mathrm{AI}}$) in the right-hand side is an element of {AI}. That is, u_i is not a member of {AI} when $s_i \neq 1$. Indeed, $(\vec{u}_1, \vec{u}_2, \ldots, \vec{u}_{d_{\mathrm{AI}}})$ is a basis set of {BS} and $X_{\mathrm{BS}} = \mathbb{Z}_{s_1} \times \mathbb{Z}_{s_2} \times \cdots \times \mathbb{Z}_{s_{d_{\mathrm{AI}}}}$ (Note that $\mathbb{Z}_1 \equiv 0$) [4]. It should be mentioned that the above discussion is not applicable to the case of superconductors, which we will see in the next chapter.

To demonstrate the power of Smith normal forms, let us perform the above processes for inversion symmetric examples. For the one-dimensional case, we find

$$(\vec{a}_1, \vec{a}_2, \vec{a}_3) = \begin{pmatrix} 1\,1\,1\,1\,0 \\ 1\,0\,1\,0\,0 \\ 0\,1\,0\,0\,0 \\ 1\,0\,0\,0\,0 \\ 0\,1\,1\,0\,1 \end{pmatrix} \begin{pmatrix} 1\,0\,0 \\ 0\,1\,0 \\ 0\,0\,1 \\ 0\,0\,0 \\ 0\,0\,0 \end{pmatrix} \begin{pmatrix} 1\,0\,0 \\ 0\,1\,0 \\ 0\,0\,1 \end{pmatrix}, \tag{2.106}$$

in which $s_1 = s_2 = s_3 = 1$. Thus, $X_{\mathrm{BS}} \simeq 0$.

For the two-dimensional case,

$$(\vec{a}_1, \vec{a}_2, \vec{a}_3, \vec{a}_4, \vec{a}_5) = \begin{pmatrix} 1\,1\,1\,1\,{-1}\,1\,0\,0\,0 \\ 1\,0\,1\,1\,{-1}\,0\,0\,0\,0 \\ 0\,1\,0\,0\;\;0\;\;0\,0\,0\,0 \\ 1\,0\,0\,1\,{-1}\,0\,0\,0\,0 \\ 0\,1\,1\,0\;\;0\;\;0\,0\,1\,0 \\ 1\,0\,1\,0\,{-1}\,0\,0\,0\,0 \\ 0\,1\,0\,1\;\;0\;\;0\,1\,0\,0 \\ 1\,0\,0\,0\;\;0\;\;0\,0\,0\,0 \\ 0\,1\,1\,1\,{-1}\,0\,0\,0\,1 \end{pmatrix} \begin{pmatrix} 1\,0\,0\,0\,0 \\ 0\,1\,0\,0\,0 \\ 0\,0\,1\,0\,0 \\ 0\,0\,0\,1\,0 \\ 0\,0\,0\,0\,2 \\ 0\,0\,0\,0\,0 \\ 0\,0\,0\,0\,0 \\ 0\,0\,0\,0\,0 \\ 0\,0\,0\,0\,0 \end{pmatrix} \begin{pmatrix} 1\,0\,0\,0\,1 \\ 0\,1\,0\,0\,0 \\ 0\,0\,1\,0\,1 \\ 0\,0\,0\,1\,1 \\ 0\,0\,0\,0\,1 \end{pmatrix}, \tag{2.107}$$

in which $s_1 = s_2 = s_3 = s_4 = 1$ and $s_5 = 2$. Thus, $X_{\mathrm{BS}} \simeq \mathbb{Z}_2$.

References

1. Po HC, Vishwanath A, Watanabe H (2017) Symmetry-based indicators of band topology in the 230 space groups. Nat Commun 8(1):50
2. Watanabe H (2019) Solid state physics, vol 54(4). AGNE Gijutsu Center
3. Watanabe H (2019) Solid state physics, vol 54(5). AGNE Gijutsu Center
4. Po HC (2020) Symmetry indicators of band topology. J Phys Condens Matter 32(26):263001
5. Bradlyn B, Elcoro L, Cano J, Vergniory MG, Wang Z, Felser C, Aroyo MI, Andrei Bernevig B (2017) Topological quantum chemistry. Nature 547(7663):298–305
6. Watanabe H (2019) Solid state physics, vol 54(7). AGNE Gijutsu Center
7. Bradley CJ, Cracknell AP (1972) The mathematical theory of symmetry in solids. Oxford University Press, Oxford
8. Teturo I, Yukito T, Yositake O (1990) Group theory and its applications in physics. Springer Series in Solid-State Sciences. Springer, Berlin

9. Dresselhaus MS, Dresselhaus G, Jorio A (2007) Group theory: application to the physics of condensed matter. Springer Science & Business Media
10. Watanabe H, Po HC, Zaletel MP, Vishwanath A (2016) Filling-enforced gaplessness in band structures of the 230 space groups. Phys Rev Lett 117:096404
11. Po HC, Watanabe H, Zaletel MP, Vishwanath A (2016) Filling-enforced quantum band insulators in spin-orbit coupled crystals. Sci Adv 2(4):e1501782
12. Watanabe H (2018) Lieb-schultz-mattis-type filling constraints in the 1651 magnetic space groups. Phys Rev B 97:165117
13. Jorrit K, de Boer J, van Wezel J, Kane Charles L, Robert-Jan S (2017) Topological classification of crystalline insulators through band structure combinatorics. Phys Rev X 7:041069
14. Elcoro L, Bradlyn B, Wang Z, Vergniory MG, Cano J, Felser C, Andrei Bernevig B, Orobengoa D, de la Flor G, Aroyo MI (2017) Double crystallographic groups and their representations on the Bilbao Crystallographic Server. J Appl Crystall 50(5):1457–1477
15. Hahn T (ed) (2006) International tables for crystallography, volume A: space-group symmetry, 5th ed. Springer, Berlin
16. Po HC, Watanabe H, Jian C-M, Zaletel MP (2017) Lattice homotopy constraints on phases of quantum magnets. Phys Rev Lett 119:127202
17. Watanabe H, Po HC, Vishwanath A (2018) Structure and topology of band structures in the 1651 magnetic space groups. Sci Adv 4(8):eaat8685
18. Po HC, Watanabe H, Vishwanath A (2018) Fragile topology and Wannier obstructions. Phys Rev Lett 121:126402
19. Bradlyn B, Wang Z, Cano J, Andrei Bernevig B (2019) Disconnected elementary band representations, fragile topology, and Wilson loops as topological indices: an example on the triangular lattice. Phys Rev B 99:045140
20. Po HC, Zou L, Vishwanath A, Senthil T (2018) Origin of mott insulating behavior and superconductivity in twisted bilayer graphene. Phys Rev X
21. Ahn J, Park S, Yang B-J (2019) Failure of Nielsen-Ninomiya theorem and fragile topology in two-dimensional systems with space-time inversion symmetry: application to twisted bilayer graphene at magic angle. Phys Rev X(9):021013

Chapter 3
Symmetry Indicators for Topological Superconductors

Bearing in mind that the symmetry-indicator theory for insulators has succeeded in extensively discovering topological insulator candidates, it is natural to think that a generalization of the theory to superconductors could facilitate the investigation of realistic topological superconductors. Since Bogoliubov quasiparticle spectra can be regarded as band structures, one might think that symmetry-indicator theory for insulators could be directly applied to Bogoliubov quasiparticle spectra. However, such a naive application has two main problems: (i) the absence of well-known symmetry-indicator formulas using $\mathbb{Z}_2$-valued Pfaffian invariants; (ii) the absence of nontrivial symmetry indicators for one-dimensional topological phases. In this chapter, we resolve these problems and establish symmetry-indicator theory for superconductors.

In Sect. 3.1, we briefly review a mean-field theoretical description of superconductors. In particular, we define pairing symmetries by symmetry-transformation properties of the superconducting order parameter and introduce symmetry representations in the superconducting phase. In Sect. 3.2, we explain the above two problems through examples. In Sect. 3.3, we discuss how to characterize Bogoliubov quasiparticle spectra by band topology, which is achieved by zero-dimensional topological invariants defined for irreducible representations of little groups. Importantly, the zero-dimensional topological invariants are defined by the difference between the target superconductor and the vacuum. In Sect. 3.4, we discuss the computational procedures of symmetry indicator groups.

S. Ono, *Topology of Superconductors*, Springer Theses,
https://doi.org/10.1007/978-981-97-7399-2_3

3.1 General Description for Superconductors

In this section, we provide a mean-field theoretical description of superconductors.

3.1.1 Bogoliubov-de Gennes Hamiltonian and Symmetry

In this thesis, we always consider superconductors that can be described by the mean-field theory. In the mean-field theory, the superconducting order parameter is described by

$$\hat{\Delta} = \frac{1}{2}\sum_{k}\sum_{\sigma\sigma'}[\Delta_{k}]_{\sigma\sigma'}\hat{c}^{\dagger}_{\sigma,k}\hat{c}^{\dagger}_{\sigma',-k} + \text{h.c.}. \tag{3.1}$$

Then, the full mean-field Hamiltonian is given by

$$\begin{aligned}\hat{H}_{\text{sc}} &= \hat{H} + \hat{\Delta} \\ &= \frac{1}{2}\left(\hat{c}^{\dagger}_{k}\ \hat{c}^{T}_{-k}\right) H^{\text{BdG}}_{k} \begin{pmatrix} \hat{c}_{k} \\ [\hat{c}^{\dagger}_{-k}]^{T} \end{pmatrix},\end{aligned} \tag{3.2}$$

where $\hat{c}^{\dagger}_{k} = (\hat{c}^{\dagger}_{1,k}, \hat{c}^{\dagger}_{2,k}, \dots, \hat{c}^{\dagger}_{N_{\text{orb}},k})$. In the following, we refer to

$$H^{\text{BdG}}_{k} = \begin{pmatrix} H_{k} & \Delta_{k} \\ \Delta^{\dagger}_{k} & -H^{*}_{-k} \end{pmatrix} \tag{3.3}$$

as Bogoliubov-de Gennes (BdG) Hamiltonian. This Hermitian matrix inherently possesses the following transformation property

$$\rho_{k}(\mathcal{C}) = \begin{pmatrix} & \xi\mathbb{1}_{N_{\text{orb}}} \\ +\mathbb{1}_{N_{\text{orb}}} & \end{pmatrix} K \tag{3.4}$$

$$\rho_{k}(\mathcal{C}) H^{\text{BdG}}_{k} = -H^{\text{BdG}}_{-k}\rho_{k}(\mathcal{C}), \tag{3.5}$$

where $\xi = \pm 1$. This is what we call *particle-hole symmetry* (PHS).[1]

Suppose that the normal phase is symmetric under a magnetic space group $\mathcal{M}$. Due to the presence of PHS $\mathcal{C}$ in the BdG Hamiltonian, the full symmetry group G is composed of the following four parts

[1] It should be noted that this PHS is different from a particle-hole transformation for $\hat{c}^{\dagger}_{R}$: $\hat{c}^{\dagger}_{R} \to \hat{c}_{R}$. This symmetry originates from Nambu spinor $\left(\hat{c}^{\dagger}_{k}\ \hat{c}^{T}_{-k}\right)$ [1].

$$\begin{aligned} G &= \mathcal{M} + \mathcal{M}\mathcal{C} \\ &= \mathcal{G} + \mathcal{A} + \mathcal{P} + \mathcal{J} \end{aligned} \tag{3.6}$$

where $\mathcal{P} = \mathcal{G}\mathcal{C}$ and $\mathcal{J} = \mathcal{A}\mathcal{C}$ are sets of particle-hole-like and chiral-like symmetries.

3.1.2 Pairing Symmetry

From Eq. (2.14), the superconducting order parameter is transformed as

$$\hat{g}\hat{\Delta}\hat{g}^{-1} = \begin{cases} \frac{1}{2}\sum_{\boldsymbol{k}}\sum_{\sigma\sigma'}\sum_{l,l'} \hat{c}^{\dagger}_{l,p_g\boldsymbol{k}}[U_{\boldsymbol{k}}(g)]_{l\sigma}[\Delta_{\boldsymbol{k}}]_{\sigma\sigma'}[U_{-\boldsymbol{k}}(g)]_{l'\sigma'}\hat{c}^{\dagger}_{l',-p_g\boldsymbol{k}} + \text{h.c.} \ (\phi_g = +1) \\ \frac{1}{2}\sum_{\boldsymbol{k}}\sum_{\sigma\sigma'}\sum_{l,l'} \hat{c}^{\dagger}_{l,-p_g\boldsymbol{k}}[U_{\boldsymbol{k}}(g)]_{l\sigma}[\Delta^*_{\boldsymbol{k}}]_{\sigma\sigma'}[U_{-\boldsymbol{k}}(g)]_{l'\sigma'}\hat{c}^{\dagger}_{l',p_g\boldsymbol{k}} + \text{h.c.} \ (\phi_g = -1) \end{cases}, \tag{3.7}$$

$$= \begin{cases} \frac{1}{2}\sum_{\boldsymbol{k}}\sum_{\sigma\sigma'} \hat{c}^{\dagger}_{\sigma,p_g\boldsymbol{k}}\left(U_{\boldsymbol{k}}(g)\Delta_{\boldsymbol{k}}U^T_{-\boldsymbol{k}}(g)\right)_{\sigma\sigma'}\hat{c}^{\dagger}_{\sigma',-p_g\boldsymbol{k}} + \text{h.c.} \ (\phi_g = +1) \\ \frac{1}{2}\sum_{\boldsymbol{k}}\sum_{\sigma\sigma'} \hat{c}^{\dagger}_{\sigma,-p_g\boldsymbol{k}}\left(U_{\boldsymbol{k}}(g)\Delta^*_{\boldsymbol{k}}U^T_{-\boldsymbol{k}}(g)\right)_{\sigma\sigma'}\hat{c}^{\dagger}_{\sigma',p_g\boldsymbol{k}} + \text{h.c.} \ (\phi_g = -1) \end{cases}, \tag{3.8}$$

where $g \in \mathcal{M}$. To satisfy $\hat{g}\hat{\Delta}\hat{g}^{-1} = \hat{\Delta}$, $\Delta_{\boldsymbol{k}}$ must be transformed as

$$U_{\boldsymbol{k}}(g)\Delta_{\boldsymbol{k}}U^T_{-\boldsymbol{k}}(g) = \chi_g\Delta_{p_g\boldsymbol{k}} \ \ (\phi_g = +1), \tag{3.9}$$

$$U_{\boldsymbol{k}}(g)\Delta^*_{\boldsymbol{k}}U^T_{-\boldsymbol{k}}(g) = \chi_g\Delta_{-p_g\boldsymbol{k}} \ \ (\phi_g = -1), \tag{3.10}$$

where $\chi_g = e^{i\theta_g} \in \mathrm{U}(1)$. The appearance of χ_g can be eliminated by the gauge transformation $\hat{c}^{\dagger}_{\sigma,\boldsymbol{k}} \to e^{-i\theta_g/2}\hat{c}^{\dagger}_{\sigma,\boldsymbol{k}}$. Then, the Nambu spinor is

$$\begin{aligned} g\left(\hat{\boldsymbol{c}}^{\dagger}_{\boldsymbol{k}} \ \hat{\boldsymbol{c}}^{T}_{-\boldsymbol{k}}\right)g^{-1} &= \left(\hat{\boldsymbol{c}}^{\dagger}_{p_g\boldsymbol{k}} \ \hat{\boldsymbol{c}}^{T}_{-p_g\boldsymbol{k}}\right)\begin{pmatrix} e^{-i\theta_g/2}\mathbb{1}_{N_{\text{orb}}} & 0 \\ 0 & e^{i\theta_g/2}\mathbb{1}_{N_{\text{orb}}} \end{pmatrix}\begin{pmatrix} U_{\boldsymbol{k}}(g) & 0 \\ 0 & U^*_{-\boldsymbol{k}}(g) \end{pmatrix} \\ &= \left(\hat{\boldsymbol{c}}^{\dagger}_{p_g\boldsymbol{k}} \ \hat{\boldsymbol{c}}^{T}_{-p_g\boldsymbol{k}}\right)\begin{pmatrix} e^{-i\theta_g/2}U_{\boldsymbol{k}}(g) & 0 \\ 0 & e^{i\theta_g/2}U^*_{-\boldsymbol{k}}(g) \end{pmatrix}. \end{aligned} \tag{3.11}$$

By redefining $\tilde{U}_{\boldsymbol{k}}(g) = e^{-i\theta_g/2}U_{\boldsymbol{k}}(g)$, the relation $\hat{g}\hat{\Delta}\hat{g}^{-1} = \hat{\Delta}$ holds. Since we can always multiply a symmetry representation by a U(1) phase factor, we define a unitary matrix of the symmetry representation of $g \in \mathcal{M}$ by

$$U^{\text{BdG}}_{\boldsymbol{k}}(g) = \begin{pmatrix} U_{\boldsymbol{k}}(g) & 0 \\ 0 & \chi_g U^*_{-\boldsymbol{k}}(g) \end{pmatrix} \tag{3.12}$$

in the superconducting phase. We define a unitary matrix for $g \in \mathcal{P}$ or $\mathcal{J}$ by

$$U^{\text{BdG}}_{\boldsymbol{k}}(g) = \begin{pmatrix} U_{\boldsymbol{k}}(g) & 0 \\ 0 & \chi_g U^*_{-\boldsymbol{k}}(g) \end{pmatrix}\begin{pmatrix} & \xi\mathbb{1}_{N_{\text{orb}}} \\ +\mathbb{1}_{N_{\text{orb}}} & \end{pmatrix} = \begin{pmatrix} & \xi U_{\boldsymbol{k}}(g) \\ \chi_g U^*_{-\boldsymbol{k}}(g) & \end{pmatrix}. \tag{3.13}$$

As a result, we have

$$\rho_{\boldsymbol{k}}(g) = \begin{cases} U_{\boldsymbol{k}}^{\mathrm{BdG}}(g) & \text{for } \phi_g = +1 \\ U_{\boldsymbol{k}}^{\mathrm{BdG}}(g)K & \text{for } \phi_g = -1 \end{cases} \tag{3.14}$$

and $\sigma_{\boldsymbol{k}}(g)$ similarly.

We introduce a map $c : G \to \mathbb{Z}_2 = \{-1, 1\}$. Here, $c_g = +1$ (-1) represents g commutes (anticommutes) with the BdG Hamiltonian $H_{\boldsymbol{k}}^{\mathrm{BdG}}$, i.e.,

$$\rho_{\boldsymbol{k}}(g)H_{\boldsymbol{k}}^{\mathrm{BdG}} = c_g H_{g\boldsymbol{k}}^{\mathrm{BdG}}\rho_{\boldsymbol{k}}(g) = \begin{cases} H_{g\boldsymbol{k}}^{\mathrm{BdG}}\rho_{\boldsymbol{k}}(g) & \text{for } c_g = +1 \\ -H_{g\boldsymbol{k}}^{\mathrm{BdG}}\rho_{\boldsymbol{k}}(g) & \text{for } c_g = -1 \end{cases}. \tag{3.15}$$

The projective factors are defined in the same way as Eqs. (2.20) and (2.26). It should be mentioned that $U_{\boldsymbol{k}}^{\mathrm{BdG}}(h)$ $(h \in \mathcal{G}_{\boldsymbol{k}})$ and $H_{\boldsymbol{k}}^{\mathrm{BdG}}$ are always block-diagonalized in the same way as Eqs. (2.29) and (2.30).

When elements of $\mathcal{M}$ do not satisfy Eq. (3.9), those symmetries are broken in the superconducting phase. Indeed, a set of $\{\chi_g\}$ for the remaining symmetries forms a one-dimensional single-valued irreducible representation of a magnetic point group. In other words, when a superconducting order parameter belongs to a multi-dimensional irreducible representation, some symmetries are broken. Once we identify the lower magnetic space group, we can apply all discussions in this thesis to the lower one. Therefore, in this thesis, we always consider that all elements in $\mathcal{M}$ are preserved.

3.2 Problems in the Original Theory

In this section, we explain why we need to improve the original symmetry-indicator theory. When we naively apply the original theory to superconductors, we face two main problems: (i) the absence of $\mathbb{Z}_2$-valued Pfaffian invariants; (ii) the absence of nontrivial X_{BS} in 1D.

3.2.1 The Absence of the Pfaffian Invariants

While normal conducting phases are characterized only by integer-valued quantities (the counts of irreducible representations of $\mathcal{G}_{\boldsymbol{k}}$), the superconducting phases are characterized by $\mathbb{Z}_2$-valued quantities in addition to integer-valued ones.

Here, we consider two-dimensional systems with PHS ($\xi = +1$) and without any spatial symmetry. According to the topological periodic table Refs. [2, 3], the $\mathbb{Z}_2$-valued Pfaffian invariant $P_{\boldsymbol{k}} \in \{0, 1\} \simeq \mathbb{Z}_2$ is well-defined at $(k_x, k_y) = (0, 0), (\pi, 0), (0, \pi)$, and (π, π). In this setting, Chern number modulo two can be computed by

$$\mathrm{Ch} = P_{(0,0)} + P_{(\pi,0)} + P_{(0,\pi)} + P_{(\pi,\pi)} \mod 2, \tag{3.16}$$

where $\mathrm{Ch} \in \mathbb{Z}$ is Chern number. This is actually a symmetry indicator for superconductors. However, the symmetry indicator group without any spatial symmetry is trivial in the original theory.

3.2.2 *The Absence of Nontrivial X_{BS} in 1D*

As mentioned in Sect. 2.6, the symmetry indicator groups X_{BS} in one-dimensional systems are always trivial. However, unlike the case of insulators, topological superconducting phases exist in one dimension [2, 3]. Thus, the original symmetry-indicator theory cannot diagnose such one-dimensional topological superconducting phases.

To see this concretely, let us discuss a prototypical example known as *Kitaev chain* [4]. The Hamiltonian is given by

$$\hat{H}_{\mathrm{Kitaev}} \equiv \sum_i \left[-\frac{t}{2}\left(\hat{c}_{i+1}^\dagger \hat{c}_i + \mathrm{h.c.}\right) - \mu \hat{c}_i^\dagger \hat{c}_i \right] + \frac{1}{2}\sum_i \left(\Delta_0 e^{i\theta} \hat{c}_i^\dagger \hat{c}_i^\dagger + \mathrm{h.c.}\right), \tag{3.17}$$

where Δ_0 is real.[2] Under the periodic boundary condition, we obtain H_k and Δ_k in Eq. (3.3) as

$$H_k = -t \cos k - \mu, \tag{3.18}$$

$$\Delta_k = i e^{i\theta} \Delta_0 \sin k. \tag{3.19}$$

The normal-phase Hamiltonian is invariant under the inversion $I : \hat{c}_k^\dagger \to \hat{c}_{-k}^\dagger$, i.e., $U_k(I) = 1$. Then, we find that $U_k(I)\Delta_k U_{-k}^T(I) = -\Delta_{-k}$; that is, $\chi_I = -1$. After computing eigenvalues of $U_k^{\mathrm{BdG}}(I) = \mathrm{diag}(1, -1)$ for eigenvectors of H_k^{BdG} with $t = 1$ and $\mu = 0$, we find that

$$\vec{b} = \left(n_{k=0}^+, n_{k=0}^-, n_{k=\pi}^+, n_{k=\pi}^-\right) = (1, 0, 0, 1), \tag{3.20}$$

which is the same as $\vec{a}_{(x=1/2,+)}$ in Eq. (2.86).

---

[2] When $\theta = 0, \pm\pi$, the system has TRS $\mathcal{T}$ ($\mathcal{T}^2 = +1$). The symmetry class is BDI, and the classification is $\mathbb{Z}$.

3.3 Abelian Group of Superconductors

In the preceding section, we have introduced two problems of the original theory. In this section, we formulate the characterization of Bogoliubov quasiparticle spectra by zero-dimensional topological invariants. In particular, we incorporate the $\mathbb{Z}_2$-nature of Pfaffian invariants into systematic calculations of the symmetry indicators for superconductors.

3.3.1 *Zero-Dimensional Topological Invariants*

In Sect. 3.2.1, we see that $\mathbb{Z}_2$-valued Pfaffian invariants can be defined at high-symmetry points. The question is how to determine which topological invariants are defined at each momentum in the presence of magnetic space-group symmetries. This can be achieved by identifying the effective internal symmetry class for $H_{\boldsymbol{k}}^{\alpha}$ in Eq. (2.30).

In a similar way to Eq. (2.23), we define a little group of G by

$$G_{\boldsymbol{k}} \equiv \left\{h \in G \mid h(\boldsymbol{k}) = \phi_h p_h \boldsymbol{k} = \boldsymbol{k} + \boldsymbol{G}\right\}. \tag{3.21}$$

Furthermore, $G_{\boldsymbol{k}}$ is also decomposed into

$$G_{\boldsymbol{k}} = \mathcal{G}_{\boldsymbol{k}} + \mathcal{A}_{\boldsymbol{k}} + \mathcal{P}_{\boldsymbol{k}} + \mathcal{J}_{\boldsymbol{k}}, \tag{3.22}$$

where $\mathcal{V}_{\boldsymbol{k}} = \{v \in \mathcal{V} | v\boldsymbol{k} = \boldsymbol{k} + \boldsymbol{G}\}$ for $\mathcal{V} = \mathcal{A}, \mathcal{P}$, and $\mathcal{J}$. Symmetries in $\mathcal{A}_{\boldsymbol{k}}, \mathcal{P}_{\boldsymbol{k}}$, and $\mathcal{J}_{\boldsymbol{k}}$ have effects on a sector $H_{\boldsymbol{k}}^{\alpha}$ in Eq. (2.30). Sometimes, they leave $H_{\boldsymbol{k}}^{\alpha}$ as it is, and other times, they transform it into a different sector. Those symmetries that do not change $H_{\boldsymbol{k}}^{\alpha}$ result in an effective internal symmetry class for each irreducible representation at $\boldsymbol{k}$. Such an internal symmetry class is called the emergent Altland-Zirnbauer symmetry class (EAZ class).

The Wigner criterion introduced in Sect. 2.3.2 is applicable to particle-hole-like symmetries. Chiral-like symmetries do not form a Kramers pair of an irreducible representation since they are unitary. What we have to know is whether they map an irreducible representation to another one or not. This is diagnosed by checking if $\{u_{\boldsymbol{k}}^{\alpha}(g)\}_{g \in \mathcal{G}_{\boldsymbol{k}}}$ and $\left\{\frac{z_{\gamma,\gamma^{-1}g\gamma}^{\boldsymbol{k}}}{z_{g,\gamma}^{\boldsymbol{k}}} u_{\boldsymbol{k}}^{\alpha}(g)\right\}_{g \in \mathcal{G}_{\boldsymbol{k}}}$ are orthogonal to each other or not, where γ is a chiral-like symmetry. Therefore, the following three quantities tell us the EAZ class for $H_{\boldsymbol{k}}^{\alpha}$ [5]:

$$W_k^\alpha(\mathcal{P}) = \frac{1}{|\mathcal{P}_k/T|} \sum_{c\in\mathcal{P}_k/T} z_{c,c}^k \mathrm{tr}[u_k^\alpha(c^2)] \in \{0, \pm 1\}, \tag{3.23}$$

$$W_k^\alpha(\mathcal{A}) = \frac{1}{|\mathcal{A}_k/T|} \sum_{a\in\mathcal{T}_k/T} z_{a,a}^k \mathrm{tr}[u_k^\alpha(a^2)] \in \{0, \pm 1\}, \tag{3.24}$$

$$W_k^\alpha(\mathcal{J}) = \frac{1}{|\mathcal{G}_k/T|} \sum_{g\in\mathcal{G}_k/T} \frac{z_{\gamma,\gamma^{-1}g\gamma}^k}{z_{g,\gamma}^k} \mathrm{tr}[u_k^\alpha(\gamma^{-1}g\gamma)]^* \mathrm{tr}[u_k^\alpha(g)] \in \{0, 1\}. \tag{3.25}$$

When $W_k^\alpha(\mathcal{V}) = 0$, additional symmetries in $\mathcal{V}_k$ transform H_k^α into another sector H_k^β. On the other hand, when $W_k^\alpha(\mathcal{V}) = \pm 1$, H_k^α is invariant under the additional symmetries. For later convenience, we introduce the following triple

$$\mathcal{W}_k[\alpha] \equiv (W_k^\alpha(\mathcal{A}), W_k^\alpha(\mathcal{P}), W_k^\alpha(\mathcal{J})). \tag{3.26}$$

Depending on the EAZ symmetry classes, the following zero-dimensional topological invariants are assigned to each sector H_k^α [6, 7] (see Table 3.1):

$$P_k^\alpha \equiv p_k^\alpha - (p_k^\alpha)^{\mathrm{vac}} \mod 2, \tag{3.27}$$

$$N_k^\alpha \equiv n_k^\alpha - (n_k^\alpha)^{\mathrm{vac}}, \tag{3.28}$$

where p_k^α is the $\mathbb{Z}_2$-valued Pfaffian invariant, and n_k^α is the number of occupied states in H_k^α. In Eqs. (3.27) and (3.28), we also introduce the Pfaffian invariant and the number of occupied bands of a reference Hamiltonian H_k^{vac}, which are defined for each sector of H_k^{vac} and denoted by $(p_k^\alpha)^{\mathrm{vac}}$ and $(n_k^\alpha)^{\mathrm{vac}}$, respectively. In practice, a reference H_k^{vac} can always be found based on H_k. For instance, since the vacuum state is intrinsically trivial, the infinite chemical potential limit of H_k is often utilized as a reference H_k^{vac}, as detailed in [7, 8]. In this thesis, we adopt this definition for the reference Hamiltonian.

To demonstrate how to perform the above processes, let us discuss two simple examples. The first example is the one-dimensional inversion symmetric case with the even-parity pairing, $\chi_I = +1$ for inversion I. As shown in Fig. 3.1, $k = 0$ and π are invariant under C and I. The little co-group is composed of $\mathcal{G}_{k=0,\pi}/T = \{e, I\}$ and $\mathcal{P}_{k=0,\pi}/T = \{C, IC\}$. Then, there are two irreducible representations $U_{k=0,\pi}^\alpha(I) = \alpha = \pm 1$, and $z_{C,C}^k = z_{IC,IC}^k = +1$. From Eq. (3.23), we find

$$W_{k=0,\pi}^\alpha(C) = \frac{1}{2}(1+1) = 1. \tag{3.29}$$

Thus, we define the Pfaffian invariants $P_k^{\alpha=\pm}$ [3] for each parity sector at $k = 0$ and π, and the band structure is characterized by $\vec{n} = (P_{k=0}^+, P_{k=0}^-, P_{k=\pi}^+, P_{k=\pi}^-)^T$. On the other hand, generic momentum $\Sigma \in (0, \pi)$ is not invariant under C and I but invariant under CI. Similarly, the little co-group is decomposed into $\mathcal{G}_\Sigma/T = \{e\}$ and $\mathcal{P}_\Sigma/T = \{IC\}$, and we find

Table 3.1 EAZ symmetry classes and their topological classification in zero dimension

EAZ	$(W_k^\alpha(\mathcal{T}), W_k^\alpha(\mathcal{P}), W_k^\alpha(\mathcal{J}))$	Classification	Topological invariant
A	(0, 0, 0)	$\mathbb{Z}$	N_k^α
AIII	(0, 0, 1)	0	None
AI	(1, 0, 0)	$\mathbb{Z}$	N_k^α
BDI	(1, 1, 1)	$\mathbb{Z}_2$	P_k^α
D	(0, 1, 0)	$\mathbb{Z}_2$	P_k^α
DIII	(−1, 1, 1)	0	None
AII	(−1, 0, 0)	$2\mathbb{Z}$	N_k^α
CII	(−1, −1, 1)	0	None
C	(0, −1, 0)	0	None
CI	(1, −1, 1)	0	None

Topological indices P_k^α and N_k^α in the table are defined by Eqs. (3.27) and (3.28)

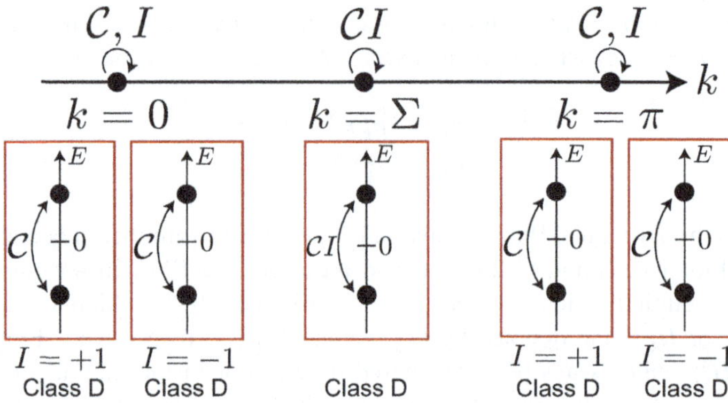

Fig. 3.1 Action of the particle-hole-like symmetries to each sector of Hamiltonians at $k = 0, \pi, \Sigma$. Here, $k = 0$ and π are invariant under PHS and the inversion, and therefore, we identify EAZ classes of sectors $I = +1$ and $I = -1$ as class D. On the other hand, $k = \Sigma$ is not invariant under them, but it is invariant under the combination $\mathcal{C}I$. Reprinted from Ref. [7] by The Author(s) licensed under CC BY 4.0

$$W_\Sigma^\alpha(\mathcal{C}) = 1. \tag{3.30}$$

As a result, we can also define the Pfaffian invariant $P_{k=\Sigma}$ at generic momenta.

The second example is one-dimensional spinful system with screw symmetry $S_{2x} : x \to x + 1/2$ with $\chi_{S_{2x}} = -1$. Since every momentum is invariant under S_{2x}, two one-dimensional irreducible representations $U_k^\alpha(S_{2z}) = i\alpha e^{-i\frac{k}{2}}$ are defined for the entire BZ. At $k = 0$ and π, the little co-group contains $\mathcal{G}_{k=0,\pi}/T = \{e, S_{2x}\}$ and $\mathcal{P}_{k=0,\pi}/T = \{\mathcal{C}, S_{2x}\mathcal{C}\}$. However, the factor systems at $k = 0$ and π are different. While $z_{\mathcal{C},\mathcal{C}}^k = +1$ at both $k = 0$ and $k = \pi$, $z_{S_{2x}\mathcal{C},S_{2x}\mathcal{C}}^{k=0} = +1$ and $z_{S_{2x}\mathcal{C},S_{2x}\mathcal{C}}^{k=\pi} = -1$. As a result, we have

$$W^{\alpha}_{k=0}(\mathcal{C}) = \frac{1}{2}(1+1) = 1, \tag{3.31}$$

$$W^{\alpha}_{k=\pi}(\mathcal{C}) = \frac{1}{2}(1-1) = 0, \tag{3.32}$$

which implies that EAZ classes for two irreducible representations at $k=0$ are class D and that those at $k=\pi$ are class A. Correspondingly, the Pfaffian invariants $P^{\alpha=\pm}_{k=0}$ are defined at $k=0$, and integer-valued invariants N^{α}_{k} are defined at $k=\pi$. Note that $N^{+}_{k} = -N^{-}_{k}$ due to PHS. Since $\mathcal{P}_{\Sigma}/T = \emptyset$ at generic momentum $\Sigma \in (0,\pi)$, the EAZ class at Σ is class A.

3.3.2 Compatibility Conditions

In Sect. 2.4, we have discussed the relations among irreducible representations at different momenta. Here, we generalize them to the zero-dimensional topological invariants. Such relations give rise to necessary conditions for zero-dimensional topological invariants to satisfy when a superconductor is gapped.

Consider $\boldsymbol{k}'$ slightly off from a higher symmetry momentum $\boldsymbol{k}$. Since topological invariants cannot change during the adiabatic deformation, zero-dimensional topological invariants at $\boldsymbol{k}$ and $\boldsymbol{k}'$ must be the same when the system is gapped. There are two possibilities:

(i) $\mathcal{W}_{\boldsymbol{k}'}[\beta] \in \{(0,1,0),(1,1,1)\}$ and there exists at least an irreducible representation such that $\mathcal{W}_{\boldsymbol{k}}[\alpha] \in \{(0,0,0),(1,0,0),(-1,0,0),(0,1,0),(1,1,1)\}$

$$P^{\beta}_{\boldsymbol{k}'} = \sum_{\alpha} c^{\alpha\beta}_{\boldsymbol{k},\boldsymbol{k}'} P^{\alpha}_{\boldsymbol{k}} + \sum_{\gamma} c^{\gamma\beta}_{\boldsymbol{k},\boldsymbol{k}'} N^{\gamma}_{\boldsymbol{k}} \mod 2, \tag{3.33}$$

where the summations are taken over α such that $\mathcal{W}_{\boldsymbol{k}}[\alpha] \in \{(0,1,0),(1,1,1)\}$ and γ such that $\mathcal{W}_{\boldsymbol{k}}[\gamma] \in \{(0,0,0),(1,0,0),(-1,0,0)\}$. Also, the coefficients $c_{\boldsymbol{k},\boldsymbol{k}'}$ is given by $U^{\alpha}_{\boldsymbol{k}}(g) = \bigoplus_{\beta} c^{\alpha\beta}_{\boldsymbol{k},\boldsymbol{k}'} U^{\beta}_{\boldsymbol{k}'}(g)$, as discussed in Sect. 2.4.
Let us go back to the discussion about the example of the above one-dimensional even-parity case. As discussed in Sect. 3.3.1, the Pfaffian invariants are defined for the entire BZ, and they must be the same when the system is fully gapped, i.e.,

$$P_{\Sigma} = P^{+}_{0} + P^{-}_{0}, \tag{3.34}$$
$$P_{\Sigma} = P^{+}_{\pi} + P^{-}_{\pi}. \tag{3.35}$$

(ii) $\mathcal{W}_{\boldsymbol{k}'}[\beta] \in \{(0,0,0),(1,0,0),(-1,0,0)\}$ and there exists at least an irreducible representation such that $\mathcal{W}_{\boldsymbol{k}}[\alpha] \in \{(0,0,0),(1,0,0),(-1,0,0)\}$

$$N_{k'}^{\beta} = \sum_{\alpha} c_{k,k'}^{\alpha\beta} N_k^{\alpha}, \tag{3.36}$$

where the summation is taken over α such that $\mathcal{W}_k[\alpha] \in \{(0,0,0), (1,0,0), (-1,0,0)\}$. To see how this happens, let us consider the example of the twofold screw symmetry case. While integer-valued invariants are defined at $k = \pi$ and $k = \Sigma$, the Pfaffian invariants are at $k = 0$. Thus, to be gapped phases, integer-valued invariants must satisfy

$$N_{\Sigma}^{\alpha} = N_{\pi}^{+}, \tag{3.37}$$

$$N_{\Sigma}^{\alpha} = N_{\pi}^{-}. \tag{3.38}$$

The reason why N_{Σ}^{α} relates to both N_{π}^{+} and N_{π}^{-} is the band connectivity originating from nonsymmorphic symmetries in Sect. 2.3.3. Since $N_{\pi}^{+} = -N_{\pi}^{-}$ due to PHS, $N_{\Sigma}^{\alpha} = 0$ when the system is gapped. It should be noted that there are no compatibility conditions between irreducible representations at $k = 0$ and $k = \Sigma$ since any map from $\mathbb{Z}_2$ to $\mathbb{Z}$ is trivial.

We refer to the first type of compatibility relations as the $\mathbb{Z}_2$-compatibility condition. After applying these equations to all high-symmetry points, we find a matrix, the counterpart of the C matrix in Sect. 2.4.

3.3.3 Solving Compatibility Conditions

In the same way as Sect. 2.4, we map a Bogoliubov quasiparticle spectrum to a set of zero-dimensional topological invariants

$$\vec{n} = (n_1, n_2, \ldots, n_D). \tag{3.39}$$

For simplicity in our approach, the entries are arranged so that the first D_P elements consist of $\mathbb{Z}_2$ numbers, while the remaining elements are integers. Note that different EAZ classes can emerge for different sectors at the same momentum. For example, at a given momentum, some sectors of irreducible representations might be associated with integer-valued invariants, whereas others could be characterized by $\mathbb{Z}_2$-valued Pfaffian invariants. Our first task is to obtain {BS} for superconductors. However, $\mathbb{Z}_2$-nature of the problem disables systematic computations of {BS}. The central idea of our approach is to introduce an auxiliary problem that is easy to handle. In the auxiliary problem, all $\mathbb{Z}_2$-valued quantities and conditions are considered as integer-valued ones.

Before moving on to discussions of general cases, it is instructive to illustrate the fundamental concepts and notations using the previously mentioned example of the one-dimensional even-parity case. As discussed in Sect. 3.3.1, the system is characterized by $\vec{n} = (P^+_{k=0}, P^-_{k=0}, P^+_{k=\pi}, P^-_{k=\pi})^T$. Compatibility conditions are $P_{k=\Sigma} = P^+_{k=0} + P^-_{k=0} = P^+_{k=\pi} + P^-_{k=\pi}$, which can be rewritten as $(1, 1, -1, -1)\vec{n} = 0 \mod 2$. To compute {BS}, forgetting about the $\mathbb{Z}_2$-nature of the Pfaffian invariants, we instead solve an alternative equation

$$P'^+_{k=0} + P'^-_{k=0} - P'^+_{k=0} + P'^-_{k=0} = 2m, \tag{3.40}$$

where the "prime" indicates that we treat it as a regular integer. At this stage, all values and relations are $\mathbb{Z}$-valued. Thus, we can easily compute the solution space of $(1, 1, -1, -1, 2)\tilde{\vec{n}} = 0$ as

$$\ker(1, 1, -1, -1, 2) = \text{span}\begin{Bmatrix} (-2\ 0\ 0\ 0\ 1)^T \\ (1\ \ 0\ 0\ 1\ 0)^T \\ (1\ \ 0\ 1\ 0\ 0)^T \\ (-1\ 1\ 0\ 0\ 0)^T \end{Bmatrix}, \tag{3.41}$$

where $\tilde{\vec{n}} = (P'^+_{k=0}, P'^-_{k=0}, P'^+_{k=0}, P'^-_{k=0}, -m)^T$. Note that, although we get four five-dimensional lists, physically meaningful parts are only the first four elements of each list. Furthermore, since the physical entries are only $\mathbb{Z}_2$-valued, we replace them with their values modulo two and get an abelian group of Bogoliubov quasiparticle spectrum as

$$\{\text{BS}\} = \text{span}\begin{Bmatrix} \vec{B}_1 = (1\ 0\ 0\ 1)^T \\ \vec{B}_2 = (1\ 0\ 1\ 0)^T \\ \vec{B}_3 = (1\ 1\ 0\ 0)^T \end{Bmatrix} \simeq (\mathbb{Z}_2)^3. \tag{3.42}$$

Here, we generalize the above computational processes. Suppose that we have d compatibility conditions encoded in $d \times D$ matrix C, where the first d_p rows out of the d are $\mathbb{Z}_2$-compatibility conditions. When the system is gapped (more precisely, gapped at least for arbitrary lines connecting any two high-symmetry points), $C\vec{n} = 0$ must hold. In the above example, $d = d_p = 1$ and $C = (1, 1, -1, -1)$. In the same way as Ref. [9], $\{\text{BS}\} \equiv \ker C \cap \left(\mathbb{Z}_2^{D_P} \times \mathbb{Z}^{D-D_P}\right)$.

A significant challenge in systematically computing {BS} arises from the $\mathbb{Z}_2$ nature of the Pfaffian invariants and compatibility conditions. In what follows, we outline the strategy to address this issue. First, we regard all rows of the C matrix as being integer-valued compatibility conditions. As mentioned above, we use such "prime" notations to indicate that we are dealing with such an auxiliary problem. For instance, the matrix C is represented as C' when the $\mathbb{Z}_2$-nature of the compatibility conditions is disregarded. To keep track of information about the $\mathbb{Z}_2$-nature of the original compatibility conditions, we introduce a new matrix $Q = (\vec{q}_1, \vec{q}_2, \ldots, \vec{q}_{d_p})$. Here, $\vec{q}_i$ is a d-dimensional vector whose j-th component is $2\delta_{ij}$, which tell us that the j-th compatibility condition in C is a $\mathbb{Z}_2$-compatibility condition.

Next, since the first d_p elements of $C'\vec{n}'$ can be nonzero and even, we consider an alternative equation

$$C'\vec{n}' = \sum_{i=1}^{d_p} m_i \vec{q}_i \quad (m_i \in \mathbb{Z}), \tag{3.43}$$

Then, we can rewrite this equation as

$$\begin{pmatrix} C' & \mathcal{Q} \end{pmatrix} \tilde{\vec{n}} = \mathbf{0}, \tag{3.44}$$

where $\tilde{\vec{n}} = (n'_1, \ldots, n'_D, -m_1, \ldots, -m_{d_p})$. Now, we have an integer-valued linear equation. It is easy to compute the null space of $\begin{pmatrix} C' & \mathcal{Q} \end{pmatrix}$, which is denoted by

$$\{\mathrm{BS}\}_f = \mathrm{span}\left(\tilde{\vec{B}}'_1, \tilde{\vec{B}}'_2, \ldots, \tilde{\vec{B}}'_{d_{\mathrm{BS}_f}}\right) \simeq \mathbb{Z}^{d_{\mathrm{BS}_f}} \tag{3.45}$$

Finally, we obtain {BS} from $\{\mathrm{BS}\}_f$. Since the physically meaningful parts are only the first D entries in $\tilde{\vec{B}}'_i$, we ignore the other elements. In addition, the first D_P entries are $\mathbb{Z}_2$-valued quantities in the original problem; thus, we replace them by their modulo two values. Then, we get

$$\{\mathrm{BS}\} = \left\{ \sum_{i=1}^{d_{\mathrm{BS}}} r_i \vec{B}_i \,\middle|\, r_i \in \mathbb{Z}_2 \text{ or } \mathbb{Z} \right\} \simeq (\mathbb{Z}_2)^n \times \mathbb{Z}^{d_{\mathrm{BS}}-n}, \tag{3.46}$$

where $\{\vec{B}_i\}_i$ is the set of all inequivalent lists $\{\tilde{\vec{B}}'_j\}_j$ after performing the above processes.

3.4 Symmetry-Indicator Theory for Superconductors

As reviewed in Sect. 2, the symmetry indicator is defined by subtracting all sets of zero-dimensional topological invariants, denoted by {AI}, from {BS}. The remaining task is to obtain {AI} and compute symmetry indicator groups, as is the case of insulators. For the latter task, we again face the same difficulty arising from the $\mathbb{Z}_2$ nature of the Pfaffian invariants, and we resolve it by alternatively introducing the auxiliary, integer-valued problem.

3.4.1 Atomic Superconductors

In Sect. 3.2.2, we have seen that the original theory could not capture one-dimensional topological superconductors. This failure is ascribed to the fact that the numbers of irreducible representations of $\mathcal{G}_k$ are not zero-dimensional topological invariants of superconductors. Recall that the zero-dimensional topological invariants are defined by the difference between a target superconductor and its vacuum Hamiltonian. Correspondingly, we also introduce such a "difference" structure to the definition of {AI}.[3]

Let $\bar{\vec{a}}_i$ and $\vec{a}_i$ be lists of the Pfaffian and the number of the appearance of irreducible representations in occupied states for $H^{\text{vac}} = \begin{pmatrix} \mathbb{1} & O \\ O & -\mathbb{1} \end{pmatrix}$ and $H^{\text{full}} = -H^{\text{vac}}$, respectively. Then, {AI} is defined by

$$\{\text{AI}\} = \left\{ \sum_i^{d_{\text{AI}}} s_i \vec{A}_i : s_i \in \mathbb{Z}_2 \text{ or } \mathbb{Z} \right\}, \tag{3.47}$$

where $\vec{A}_i = \vec{a}_i - \bar{\vec{a}}_i$. To see whether this works well, let us apply the above discussions to the Kitaev chain with inversion symmetry. A similar demonstration has already been performed in Refs. [8, 11], but we here discuss it based on our notation. The BdG Hamiltonian and the representation are given in Sect. 3.2.2. For this model, we get

$$\vec{b}^{\text{BdG}} = (n_0^+, n_0^-, n_\pi^+, n_\pi^-) = (1, 0, 0, 1), \tag{3.48}$$

$$\vec{b}^{\text{vac}} = (0, 1, 0, 1). \tag{3.49}$$

Therefore, we find

$$\vec{B} = \vec{b}^{\text{BdG}} - \vec{b}^{\text{vac}} = (N_0^+, N_0^-, N_\pi^+, N_\pi^-) = (1, 0, 0, 1) - (0, 1, 0, 1) = (1, -1, 0, 0). \tag{3.50}$$

On the other hand, there are three independent sets of trivial atomic limits H^{vac} and H^{full} that correspond to

$$\vec{a}_1 = (1, 0, 1, 0), \quad \vec{a}_2 = (0, 1, 0, 1), \quad \vec{a}_3 = (1, 0, 0, 1); \tag{3.51}$$

$$\bar{\vec{a}}_1 = (0, 1, 0, 1), \quad \bar{\vec{a}}_2 = (1, 0, 1, 0), \quad \bar{\vec{a}}_3 = (0, 1, 1, 0), \tag{3.52}$$

which are obtained in Sect. 2.5. We have

---

[3] The reason why such a difference structure appears can be understood by revisiting the definition of topologically trivial superconductors, as discussed in Ref. [8, 10]. In Ref. [10], we say a superconductor is trivial if there exists a smooth deformation

$$H_k^{\text{BdG}} \oplus \begin{pmatrix} -\mathbb{1}_{N'_{\text{orb}}} & \\ & +\mathbb{1}_{N'_{\text{orb}}} \end{pmatrix} \oplus \begin{pmatrix} +\mathbb{1}_{N''_{\text{orb}}} & \\ & -\mathbb{1}_{N''_{\text{orb}}} \end{pmatrix} \sim H^{\text{vac}} \oplus \begin{pmatrix} +\mathbb{1}_{N'_{\text{orb}}} & \\ & -\mathbb{1}_{N'_{\text{orb}}} \end{pmatrix} \oplus \begin{pmatrix} +\mathbb{1}_{N''_{\text{orb}}} & \\ & -\mathbb{1}_{N''_{\text{orb}}} \end{pmatrix}.$$

Readers interested in more details may refer to Ref. [10].

$$\vec{A}_1 = \vec{a}_1 - \bar{\vec{a}}_1 = (1, 0, 1, 0) - (0, 1, 0, 1) = (1, -1, 1, -1), \tag{3.53}$$

$$\vec{A}_2 = \vec{a}_2 - \bar{\vec{a}}_2 = (0, 1, 0, 1) - (1, 0, 1, 0) = (-1, 1, -1, 1), \tag{3.54}$$

$$\vec{A}_3 = \vec{a}_3 - \bar{\vec{a}}_3 = (1, 0, 0, 1) - (0, 1, 1, 0) = (1, -1, -1, 1). \tag{3.55}$$

Since $\vec{A}_2 = -\vec{A}_1$, we have

$$\{\text{AI}\} = \text{span}\left(\vec{A}_1, \vec{A}_3\right) \simeq \mathbb{Z}^2. \tag{3.56}$$

We see that

$$\vec{B} = \vec{b}^{\text{BdG}} - \vec{b}^{\text{vac}} = \frac{1}{2}\vec{A}_1 + \frac{1}{2}\vec{A}_3 \notin \{\text{AI}\}. \tag{3.57}$$

The fractional coefficients imply the nontrivial topology of H_k^{BdG}. Therefore, the refined method can detect the Kitaev chain.

3.4.2 *Symmetry Indicator Group for Superconductors*

Similar to the case of solving compatibility conditions, we define lists $\vec{P}_i$ whose j-th component is $2\delta_{ij}$ $(1 \leq i \leq D_P, 1 \leq j \leq D)$, which allows us to keep track of the $\mathbb{Z}_2$-ness of Pfaffian invariants. These lists $\{\vec{P}_i\}_{i=1}^{D_P}$ form an abelian group $(2\mathbb{Z})^{D_P}$ denoted by P, i.e.,

$$P = \text{span}\left(\vec{P}_1, \vec{P}_2, \ldots, \vec{P}_{D_P}\right) \simeq (2\mathbb{Z})^{D_P}. \tag{3.58}$$

Now, we are ready to define our auxiliary problem. We construct a free abelian group from $\{\text{BS}\}$ as

$$\{\text{BS}\}_f = \text{span}\left(\tilde{\vec{B}}_1, \tilde{\vec{B}}_2, \ldots, \tilde{\vec{B}}_{d_{\text{BS}_f}}\right) \simeq \mathbb{Z}^{d_{\text{BS}_f}}, \tag{3.59}$$

where $\tilde{\vec{B}}_i = \sum_{j=1}^{d_{\text{BS}}} r'_{ij}\vec{B}'_j + \sum_{j=1}^{D_P} m_{ij}\vec{P}_j$ with $r'_{ij}, m_{ij} \in \mathbb{Z}$. It should be noted that the prime indicates that we are considering the auxiliary problem. By construction, P is always a subgroup of $\{\text{BS}\}_f$ and $\{\text{BS}\} = \{\text{BS}\}_f/P$. Similarly, we construct

$$\{\text{AI}\}_f = \text{span}\left(\tilde{\vec{A}}_1, \tilde{\vec{A}}_2, \ldots, \tilde{\vec{A}}_{d_{\text{AI}_f}}\right) \simeq \mathbb{Z}^{d_{\text{AI}_f}}, \tag{3.60}$$

where $\tilde{\vec{A}}_i = \sum_{j=1}^{d_{\text{AI}}} s'_{ij}\vec{A}'_j + \sum_{j=1}^{D_P} l_{ij}\vec{P}_j$ with $s'_{ij}, l_{ij} \in \mathbb{Z}$. Again, $P \subseteq \{\text{AI}\}_f$ and $\{\text{AI}\} = \{\text{AI}\}_f/P$.

Since $\{\mathrm{AI}\}_f \subseteq \{\mathrm{BS}\}_f$, we find $P \subseteq \{\mathrm{AI}\}_f \subseteq \{\mathrm{BS}\}_f$. Therefore, we can apply the third isomorphism theorem [12] to X_{BdG}, and we get

$$X_{\mathrm{BdG}} = \frac{\{\mathrm{BS}\}}{\{\mathrm{AI}\}} = \frac{\{\mathrm{BS}\}_f/P}{\{\mathrm{AI}\}_f/P} = \frac{\{\mathrm{BS}\}_f}{\{\mathrm{AI}\}_f} = X_f. \tag{3.61}$$

The original problem with the $\mathbb{Z}_2$-nature is finally reduced to the problem of computing a quotient group between free abelian groups, which can be easily handled. Note that X_f is always a finite abelian group, as one can verify that the dimensions of $\{\mathrm{BS}\}_f$ and $\{\mathrm{AI}\}_f$, d_{BS_f} and d_{AI_f}, are always the same for all symmetry settings from explicit calculations.

Using the technique developed here, Ref. [7] compute X_f for one-dimensional irreducible representations of the point groups in magnetic space groups. We also take $\xi = [\rho(\mathcal{C})]^2 = \pm 1$ and the spinful/spinless nature of the system into account. Also, for several symmetry settings, Refs. [6–8, 10, 11] discussed what kinds of topological superconducting phases are indicated by symmetry indicator groups. However, while the indications of symmetry indicators for insulators have been completely understood [13–16], those for superconductors have not yet been established.

3.5 Conclusions of This Chapter

In this chapter, we have generalized the theory of symmetry indicators to superconductors by developing the computation techniques for the $\mathbb{Z}_2$-nature of Pfaffian invariants and by revisiting the definition of trivial superconductors. Importantly, our refined symmetry-indicator theory can detect one-dimensional topological superconducting phases in various symmetry settings. As a result, our refinement proposed in this chapter can capture topological phases with zero-dimensional Majorana modes, which belong to the same phases as those constructed by one-dimensional topological superconducting phases. Surprisingly, the refinement enables us to diagnose not only the higher-order topological superconducting phases but also the first-order topological superconducting phases in two spatial dimensions without inversion symmetry, as shown in Sect. 3.4.

There are several important questions for future studies. One is to study what kinds of topological superconducting phases could be detected by symmetry indicators. Although we have developed the method to systematically compute symmetry indicator groups and have derived topological indices in several symmetry settings, we do not understand the indications of symmetry indicators in almost all symmetry settings. The authors of Ref. [13–18] have established the relationships between symmetry indicators and topological insulators or semimetals by comparing symmetry indicators with topological classifications or invariants. Having in mind the success of topological material investigations based on symmetry indicators [19–22], we should derive a complete understanding of the relations between our refined symmetry indicators and topological/nodal superconducting phases. We leave it as a promising future study.

Another interesting question is how our theory could be applied to surface superconductivity, especially for the anomalous surface states arising from a topological bulk [23]. Recently, surface topological superconductivity in iron-based superconductors has attracted much attention [24–29]. Given that an enormous number of topological insulator candidates exist [19–21], it is crucial to study an efficient diagnostic scheme for surface topological superconductivity.

References

1. Chiu C-K, Teo JCY, Schnyder AP, Ryu S (2016) Classification of topological quantum matter with symmetries. Rev Mod Phys 88:035005
2. Kitaev A (2009) Periodic table for topological insulators and superconductors. AIP Conf Proc 1134(1):22–30
3. Ryu S, Schnyder AP, Furusaki A, Ludwig AWW (2010) Topological insulators and superconductors: tenfold way and dimensional hierarchy. New J Phys 12(6):065010
4. Kitaev AYu (2001) Unpaired Majorana fermions in quantum wires. Physics-Uspekhi 44(10S):131–136
5. Shiozaki K, Sato M, Gomi K (2022) Atiyah-Hirzebruch spectral sequence in band topology: general formalism and topological invariants for 230 space groups. Phys Rev B 106:165103 Oct
6. Geier M, Brouwer PW, Trifunovic L (2020) Symmetry-based indicators for topological Bogoliubov-de Gennes Hamiltonians. Phys Rev B 101:245128 Jun
7. Ono S, Po HC, Shiozaki K (2021) $\mathbb{Z}_2$-enriched symmetry indicators for topological superconductors in the 1651 magnetic space groups. Phys Rev Res 3:023086
8. Skurativska A, Neupert T, Fischer MH (2020) Atomic limit and inversion-symmetry indicators for topological superconductors. Phys Rev Res 2:013064 Jan
9. Po HC, Vishwanath A, Watanabe H (2017) Symmetry-based indicators of band topology in the 230 space groups. Nat Commun 8(1):50
10. Ono S, Po HC, Watanabe H (2020) Refined symmetry indicators for topological superconductors in all space groups. Sci Adv 6(18):eaaz8367
11. Shiozaki K (2019) Variants of the symmetry-based indicator
12. David S (1998) Dummit and Richard M. Wiley, Foote. Abstract Algebra
13. Song Z, Zhang T, Fang C (2018) Diagnosis for nonmagnetic topological semimetals in the absence of spin-orbital coupling. Phys Rev X 8:031069 Sep
14. Khalaf E, Po HC, Vishwanath A, Watanabe H (2018) Symmetry indicators and anomalous surface states of topological crystalline insulators. Phys Rev X 8:031070
15. Song Z, Zhang T, Fang Z, Fang C (2018) Quantitative mappings between symmetry and topology in solids. Nat Commun 9(1):3530
16. Peng B, Jiang Y, Fang Z, Weng H, Fang C (2022) Topological classification and diagnosis in magnetically ordered electronic materials. Phys Rev B 105:235138 Jun
17. Ono S, Watanabe H (2018) Unified understanding of symmetry indicators for all internal symmetry classes. Phys Rev B 98:115150 Sep
18. Elcoro L, Wieder BJ, Song Z, Xu Y, Bradlyn B, Andrei Bernevig B (2021) Magnetic topological quantum chemistry. Nat Commun 12(1):5965
19. Zhang T, Jiang Y, Song Z, Huang H, He Y, Fang Z, Weng H, Fang C (2019) Catalogue of topological electronic materials. Nature 566(7745):475–479

20. Vergniory MG, Elcoro L, Felser C, Regnault N, Andrei Bernevig B, Wang Z (2019) A complete catalogue of high-quality topological materials. Nature 566(7745):480–485
21. Tang F, Po HC, Vishwanath A, Wan X (2019) Comprehensive search for topological materials using symmetry indicators. Nature 566(7745):486–489
22. Xu Y, Elcoro L, Song Z-D, Wieder BJ, Vergniory MG, Regnault N, Chen Y, Felser C, Andrei Bernevig B (2020) High-throughput calculations of magnetic topological materials. Nature 586(7831):702–707
23. Fu L, Kane CL (2008) Superconducting proximity effect and Majorana Fermions at the surface of a topological insulator. Phys Rev Lett 100:096407
24. Zhang P, Yaji K, Hashimoto T, Ota Y, Kondo T, Okazaki K, Wang Z, Wen J, Gu GD, Ding H, Shin S (2018) Observation of topological superconductivity on the surface of an iron-based superconductor. Science 360(6385):182–186
25. Liu Q, Chen C, Zhang T, Peng R, Yan Y-J, Wen C-H-P, Lou X, Huang Y-L, Tian J-P, Dong X-L, Wang G-W, Bao W-C, Wang Q-H, Yin Z-P, Zhao Z-X, Feng D-L (2018) Robust and clean Majorana Zero mode in the vortex core of high-temperature superconductor $(Li_{0.84}Fe_{0.16})OHFeSe$. Phys Rev X 8:041056
26. Kong L, Zhu S, Papaj M, Hui Chen L, Cao HI, Xing Y, Liu W, Wang D, Fan P, Sun Y, Shixuan D, Schneeloch J, Zhong R, Genda G, Liang F, Gao H-J, Ding H (2019) Half-integer level shift of vortex bound states in an iron-based superconductor. Nat Phys 15(11):1181–1187
27. Machida T, Sun Y, Pyon S, Takeda S, Kohsaka Y, Hanaguri T, Sasagawa T, Tamegai T (2019) Zero-energy vortex bound state in the superconducting topological surface state of Fe(Se, Te). Nat Mater 18(8):811–815
28. Wang Z, Rodriguez JO, Jiao L, Howard S, Graham M, Gu GD, Hughes TL, Morr DK, Madhavan V (2020) Evidence for dispersing 1D Majorana channels in an iron-based superconductor. Science 367(6473):104–108
29. Chiu C-K, Machida T, Huang Y, Hanaguri T, Zhang F-C (2020) Scalable Majorana vortex modes in iron-based superconductors. Sci Adv 6(9):eaay0443

Chapter 4
Classifications of Superconducting Nodes

In Sect. 1.2.2, we have offered a broad outline of our concepts for classifying nodes on line segments in momentum space. In this chapter, we discuss a unified approach to classifying superconducting nodes that are fixed on the lines.

This chapter is organized as follows. In Sect. 4.1, we introduce some notions in Atiyah-Hirzebruch spectral sequence, which are used to formulate our theory. Also, we show how to diagnose dimensionality of a gapless point on a line based on compatibility conditions. In Sect. 4.2, we discuss a systematic classification of point nodes on the lines in the presence of point group symmetries, which compensate for insufficiency of compatibility conditions. In Sect. 4.3, we integrate the point-node classifications into the symmetry-based analysis to classify gapless points pinned to the lines. In Sect. 4.4, we comment on superconducting nodes that are out of the scope of this section. More concretely, we discuss the relationship among such superconducting nodes, symmetry indicators, and Atiyah-Hirzebruch spectral sequence, which would be a solution for a complete classification of topologically stable nodes. Several details are discussed in appendices to avoid digressing from the main subjects.

4.1 Formalism

In this section, we outline several fundamental components essential to our systematic classification. We explain how to use them in Sects. 4.2 and 4.3.

4.1.1 Cell Decomposition

In this subsection, we discuss the *cell decomposition* of momentum space [2]. In our framework introduced below, we divide BZ into points, line segments, polygons,

S. Ono, *Topology of Superconductors*, Springer Theses,
https://doi.org/10.1007/978-981-97-7399-2_4

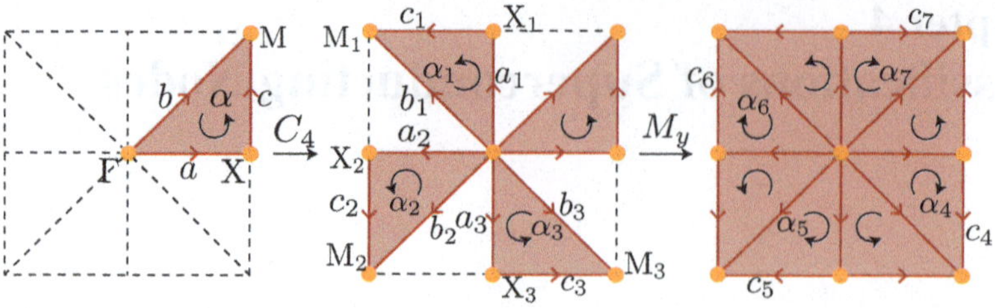

Fig. 4.1 Cell decomposition for $p4mm$. For the cell decomposition of $p4mm$, we first identify a fundamental domain of BZ and decompose it into 0-cells (orange circles), 1-cells (red lines), and a 2-cell (pink polygon), as shown in the left panel. The orientations of 1-cells and 2-cells are indicated by red and black arrows, respectively. Next, each p-cell within the fundamental domain is rotated according to the fourfold rotation symmetry. Finally, by applying mirror symmetry, we obtain the cell decomposition depicted in the right panel. Reprinted from Ref. [1] by The Author(s) licensed under CC BY 4.0

and polyhedrons, which are called 0-cells, 1-cells, 2-cells, and 3-cells, respectively. Before moving on to discussions for general cases, we start with wallpaper group $p4mm$ as an example.

Here, we derive the cell decomposition shown in the right panel of Fig. 4.1. In this figure, orange circles, solid red lines, and pink polygons denote 0-cells, 1-cells, and 2-cells, respectively. First, we find a fundamental domain of BZ and decompose the fundamental domain into three 0-cells, three 1-cells, and one 2-cell, as shown in the left panel of Fig. 4.1. Next, we apply symmetry operations to this fundamental domain, which results in the cell decomposition of the entire BZ:

$$\mathcal{C}_0 = \{\Gamma, \mathrm{X}, \mathrm{M}, \mathrm{X}_1, \mathrm{M}_1, \mathrm{X}_2, \mathrm{M}_2, \mathrm{X}_3, \mathrm{M}_3\}, \tag{4.1}$$

$$\mathcal{C}_1 = \{a, b, c, a_1, b_1, c_1, a_2, b_2, c_2, a_3, b_3, c_3, c_4, c_5, c_6, c_7\}, \tag{4.2}$$

$$\mathcal{C}_2 = \{\alpha, \alpha_1, \alpha_2, \alpha_3, \alpha_4, \alpha_5, \alpha_6, \alpha_7\}, \tag{4.3}$$

where $\mathcal{C}_p$ $(p = 0, 1, 2)$ represents a set of p-cells. It should be noted that, although some p-cells are related to other p-cells by symmetries and reciprocal lattice vectors, we here label them with different symbols. For instance, $\mathrm{X}_2 = (-\pi, 0)$ and $\mathrm{X}_1 = (0, \pi)$ are related to X.

We now describe a methodology to obtain a cell decomposition for an arbitrary symmetry setting. Following the procedure illustrated above, we first identify a fundamental domain of BZ and then partition this domain into a set of p-cells $\{D_i^p\}_i$ for $p = 0, 1, \ldots, d$. Next, we apply symmetries to the decomposition of the fundamental domain. This yields the complete set of p-cells in BZ

$$\mathcal{C}_p \equiv \bigcup_i \bigcup_{g \in G/T} D^p_{g(i)}, \tag{4.4}$$

where $D^p_{g(i)} = gD^p_i$. As mentioned above, some p-cells are equivalent or symmetry-related to others in this construction. Here, we do not take the relations among p-cells into account, which will be considered in the construction of E_1-pages in Sect. 4.1.2. Each p-cell satisfies the following conditions:

(i) Any two p-cells in $\mathcal{C}_p$ do not overlap, i.e., $D^p_i \cap D^p_j = \emptyset$ $(i \neq j)$.
(ii) Let $\boldsymbol{k}$ be a point in a p-cell D^p_i. Then, $g\boldsymbol{k} = \boldsymbol{k} +^{\exists} \boldsymbol{G}$ or $g\boldsymbol{k} \in D^p_{g(i)}$.
(iii) The boundary ∂D^p_i is composed of $(p-1)$-cells except for $p = 0$.
(iv) An orientation is assigned to each p-cell ($p \geq 1$) in a symmetric manner.
(v) No two boundary p-cells of a $(p+1)$-cell are equivalent or related by symmetry.

For our purpose of systematically classifying gapless points pinned to lines, condition (v) is essential. In Appendix 1, we provide our decomposition for each type of lattices.

4.1.2 E_1-Pages

As discussed in Sect. 3.3.1, we determine EAZ class for each irreducible representation by computing the Wigner criteria in Eqs. (3.23)–(3.25). In our cell decomposition, the little groups $G_{\boldsymbol{k}} = \mathcal{G}_{\boldsymbol{k}} + \mathcal{A}_{\boldsymbol{k}} + \mathcal{P}_{\boldsymbol{k}} + \mathcal{J}_{\boldsymbol{k}}$ at any point $\boldsymbol{k}$ in a p-cell D^p are in common, which are hereafter denoted by $G_{D^p} = \mathcal{G}_{D^p} + \mathcal{A}_{D^p} + \mathcal{P}_{D^p} + \mathcal{J}_{D^p}$ for simplicity. Thus, EAZ classes are also the same and denoted by $\mathcal{W}_{D^p_i}[\alpha]$. Then, it is enough for our purpose to consider a point $\boldsymbol{k}$ in D^p. In the following, we define abelian groups $E_1^{p,0}$ based on EAZ classes for p-cells, which can be interpreted as the classification of zero-dimensional topological phases $\{H^\alpha_{\boldsymbol{k}}\}_{\alpha,\boldsymbol{k}}$.

The set $\mathcal{C}_p$ of p-cells is composed of N_p subsets defined by $S_{D^p_i} = \{D^p_{g(i)} = gD^p_i \mid g \in G\}$, where D^p_i is a representative p-cell in $S_{D^p_i}$. A set of independent p-cells is given by

$$F^p \equiv \{D^p_i\}_{i=1}^{N_p}. \tag{4.5}$$

In Ref. [2], it is shown that $E_1^{p,0}$ is the direct sum of the classification of zero-dimensional topological phases of $\{H^\alpha_{\boldsymbol{k}}\}_\alpha$ at a point $\boldsymbol{k}$ in each $D^p_i \in F^p$. Then, $E_1^{p,0}$ is completely determined by $\mathcal{W}_{D^p_i}[\alpha]$ for each irreducible representation and each $D^p_i \in F^p$. In other words, $E_1^{p,0}$ is defined by

$$E_1^{p,0} \equiv \bigoplus_{i|_{D^p_i \in F^p}} \left(\mathbb{Z}_2^{\oplus_\alpha} \oplus \mathbb{Z}^{\oplus_\beta}\right), \tag{4.6}$$

where we perform the summation in the following way:

(a) $\mathcal{W}_{D^p_i}[\alpha] \in \{(0, 1, 0), (1, 1, 1)\}$ and $\mathcal{W}_{D^p_i}[\beta] \in \{(0, 0, 0), (1, 0, 0), (-1, 0, 0)\}$ (see Eq. (3.26));

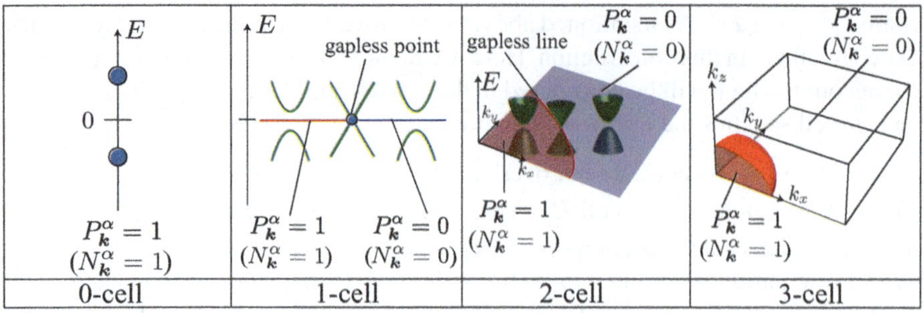

Fig. 4.2 Elements of $E_1^{p,0}$. For $p=0$, the elements correspond to gapped states at 0-cells. When $p \geq 1$, there exist two p-dimensional regions characterized by different zero-dimensional topological invariants. Since the topological invariants must be the same for gapped regions, this difference leads to the emergence of gapless states along the boundaries of the p-cells. Reprinted from Ref. [1] by The Author(s) licensed under CC BY 4.0

(b) When an irreducible representation on D_i^p is transformed to others by antiunitary or chiral-like symmetries, we consider only one representative of these irreducible representations on D_i^p.

For $p \geq 1$, $E_1^{p,0}$ denotes a set of gapless states with $(p-1)$-dimensional gapless domain walls on p-cells. This is intuitively grasped as changes in the zero-dimensional topological invariants defined on p-cells, as illustrated in Fig. 4.2. Consider a 1-cell as an example. Here, we can define an identical set of zero-dimensional topological invariants at every point in this 1-cell. However, they need not be the same values at all points along the 1-cell. When momentum is regarded as a deformation parameter, the presence of different zero-dimensional topological invariants on the 1-cell mandates the existence of gapless points on this 1-cell, as depicted in Fig. 4.2. The potential changes in zero-dimensional topological invariants on a 1-cell correspond to the classifications of zero-dimensional topological phases of $\{H_{\boldsymbol{k}}^{\alpha}\}_{\alpha}$ at a given point $\boldsymbol{k}$ on the 1-cell. This interpretation aligns with the previously mentioned $E_1^{1,0}$. In a similar manner, $E_1^{2,0}$ and $E_1^{3,0}$ are also interpreted as representing gapless lines and surfaces on 2- and 3-cells, respectively, as shown in Fig. 4.2. It is important to note that the gapless points and lines associated with $E_1^{1,0}$ and $E_1^{2,0}$ are not always genuine point and line nodes. In some cases, these nodes may form part of higher-dimensional superconducting nodes.

Based on this gapless interpretation, we can characterize any system by a list of $\mathbb{Z}_2$- and $\mathbb{Z}$-valued quantities (called *band labels* in this thesis)

$$\mathfrak{n}^{(p)} = (\mathfrak{P}_{D_1^p}^{\alpha_1}, \mathfrak{P}_{D_1^p}^{\alpha_2}, \dots \mathfrak{N}_{D_1^p}^{\beta_1}, \dots, \mathfrak{P}_{D_2^p}^{\alpha_1'}, \dots, \mathfrak{N}_{D_2^p}^{\beta_1'}, \dots), \tag{4.7}$$

where $\mathfrak{P}_{D_i^p}^{\alpha}$ and $\mathfrak{N}_{D_i^p}^{\beta}$ are $\mathbb{Z}_2$-valued and $\mathbb{Z}$-valued band labels, respectively. While band labels for $p=0$ are exactly the zero-dimensional topological invariants in Eqs. (3.27) and (3.28), those for p-cells ($p \geq 1$) represent changes of the zero-dimensional topological invariants. Using these band labels, we formulate the abelian group $E_1^{p,0}$ as

$$E_1^{p,0} = \bigoplus_{i|D_i^p \in F^p} \left(\bigoplus_{\alpha} \mathbb{Z}_2[\mathfrak{b}^{(p)}_{D_i^p,\alpha}] \oplus \bigoplus_{\beta} \mathbb{Z}[\mathfrak{b}^{(p)}_{D_i^p,\beta}] \right), \tag{4.8}$$

where $\{\mathfrak{b}^{(p)}_{D_i^p,\alpha}\}$ represent generators of $E_1^{p,0}$ by which an arbitrary $\mathfrak{n}^{(p)}$ can be expanded, and the summation over α and β are take in the same way as in Eq. (4.6). Furthermore, $\mathbb{Z}_2[\mathfrak{b}^{(p)}_{D_i^p,\alpha}]$ and $\mathbb{Z}[\mathfrak{b}^{(p)}_{D_i^p,\beta}]$ denote abelian groups generated by $\mathfrak{b}^{(p)}_{D_i^p,\alpha}$ and $\mathfrak{b}^{(p)}_{D_i^p,\beta}$.

Here, we explain how to construct the generator $\mathfrak{b}^{(p)}_{D_i^p,\alpha}$. Let $U^{\alpha}_{D_i^p}$ denote an irreducible representation at a point $\boldsymbol{k}$ inside on a p-cell D_i^p in $\mathcal{C}_p$. Each $\mathfrak{b}^{(p)}_{D_i^p,\alpha}$ is generated by an irreducible representation $U^{\alpha}_{D_i^p}$. Suppose that we have a nontrivial band label $\mathfrak{P}^{\alpha}_{D_i^p} = 1$ or $\mathfrak{N}^{\alpha}_{D_i^p} = 1$ for $U^{\alpha}_{D_i^p}$. Band labels on p-cells related to D_i^p are determined as follows. As mentioned in Sect. 2.3.1, irreducible representations $U^{\alpha'}_{D^p_{g(i)}}$ and $U^{\alpha}_{D_i^p}$ are related as

$$U^{\alpha'}_{D^p_{g(i)}}(h') = \begin{cases} \frac{z_{h',g}}{z_{g,g^{-1}h'g}} W_{\boldsymbol{k}n}(g) U^{\alpha}_{D_i^p}(g^{-1}h'g) W^{\dagger}_{\boldsymbol{k}n}(g) & \text{for } \phi_g = +1 \\ \frac{z_{h',g}}{z_{g,g^{-1}h'g}} W_{\boldsymbol{k}n}(g) [U^{\alpha}_{D_i^p}(g^{-1}h'g)]^* W^{\dagger}_{\boldsymbol{k}n}(g) & \text{for } \phi_g = -1 \end{cases}, \tag{4.9}$$

where $g \in G$ and $h' \in \mathcal{G}_{D^p_{g(i)}}$.[1] Correspondingly, band labels on $D^p_{g(i)}$ must satisfy

$$\mathfrak{P}^{\alpha'}_{D^p_{g(i)}} = \mathfrak{P}^{\alpha}_{D_i^p}, \tag{4.10}$$

$$\mathfrak{N}^{\alpha'}_{D^p_{g(i)}} = \begin{cases} \mathfrak{N}^{\alpha}_{D_i^p} & \text{for } c_g = +1 \\ -\mathfrak{N}^{\alpha}_{D_i^p} & \text{for } c_g = -1 \end{cases}. \tag{4.11}$$

As a result, we have a set of band labels such that only $\mathfrak{P}^{\alpha}_{D_i^p}$ ($\mathfrak{N}^{\alpha}_{D_i^p}$) and symmetry-related band labels are 1 (1 or -1 for EAZ class A and AI; 2 or -2 for EAZ class AII), which is exactly $\mathfrak{b}^{(p)}_{D_i^p,\alpha}$. It should be noted that we have included the symmetry relations among p-cells in the basis.

For concreteness, let us consider a one-dimensional system with an even-parity pairing in class D. A fundamental domain in Fig. 4.3a is decomposed into two 0-cells (labeled by Γ and X) and one 1-cell a. By applying the inversion symmetry I to the fundamental domain, we have the following cell decomposition:

[1] Since $U^{\alpha}_{D_i^p}$ is defined at a point $\boldsymbol{k}$, we use $W_{\boldsymbol{k}n}(g)$ defined in Eq. (2.54) as it is.

$$\mathcal{C}_0 \equiv \{\Gamma, \mathrm{X}, \mathrm{X}' = I\mathrm{X}\}, \tag{4.12}$$
$$\mathcal{C}_1 \equiv \{a, a' = Ia\}, \tag{4.13}$$

where $F^0 = \{\Gamma, \mathrm{X}\}$ and $F^1 = \{a\}$. We then proceed to determine the classifications for each irreducible representation at the 0-cells and 1-cells. Figure 4.3a demonstrates how particle-hole-like symmetries affect each sector of the Hamiltonians at each cell in F^p. Based on this analysis, it becomes evident that the EAZ classes at the Γ and X are class D, and similarly, the EAZ class at point a is also identified as class D. Therefore, we find that $E_1^{0,0} = (\mathbb{Z}_2)^4$ and $E_1^{1,0} = \mathbb{Z}_2$.

Next, we construct generators of E_1-pages in Eq. (4.8). Since EAZ classes are class D for each parity at the 0-cells, we define the Pfaffian invariants $\mathfrak{P}^{\alpha=\pm}_{D^0 \in \mathcal{C}_0}$ [14], and they form a list of band labels $(\mathfrak{P}^+_\Gamma, \mathfrak{P}^-_\Gamma, \mathfrak{P}^+_\mathrm{X}, \mathfrak{P}^-_\mathrm{X}, \mathfrak{P}'^+_\mathrm{X}, \mathfrak{P}'^-_\mathrm{X})$. Similarly, since EAZ classes of the 1-cells are also class D, we can also define the Pfaffian invariant on the 1-cells a and a'. Again, we have a list of band labels $(\mathfrak{P}_a, \mathfrak{P}_{a'})$. We then obtain the generators of $E_1^{0,0}$ and $E_1^{1,0}$ by following the above discussions. Since X′ and X are equivalent and a' is symmetry-related to a, $\mathfrak{P}'^\pm_\mathrm{X} = \mathfrak{P}^\pm_\mathrm{X}$ and $\mathfrak{P}_{a'} = \mathfrak{P}_a$. Therefore, we obtain a basis set

$$\mathfrak{b}^{(0)}_{\Gamma,+} = (1, 0, 0, 0, 0, 0); \tag{4.14}$$
$$\mathfrak{b}^{(0)}_{\Gamma,-} = (0, 1, 0, 0, 0, 0); \tag{4.15}$$
$$\mathfrak{b}^{(0)}_{\mathrm{X},+} = (0, 0, 1, 0, 1, 0); \tag{4.16}$$
$$\mathfrak{b}^{(0)}_{\mathrm{X},-} = (0, 0, 0, 1, 0, 1); \tag{4.17}$$
$$\mathfrak{b}^{(1)}_{a} = (1, 1), \tag{4.18}$$

and $E_1^{0,0} = \mathbb{Z}_2^4$ and $E_1^{1,0} = \mathbb{Z}_2$ are generated as

$$E_1^{0,0} = \mathbb{Z}_2[\mathfrak{b}^{(0)}_{\Gamma,+}] \oplus \mathbb{Z}_2[\mathfrak{b}^{(0)}_{\Gamma,-}] \oplus \mathbb{Z}_2[\mathfrak{b}^{(0)}_{\mathrm{X},+}] \oplus \mathbb{Z}_2[\mathfrak{b}^{(0)}_{\mathrm{X},-}], \tag{4.19}$$
$$E_1^{1,0} = \mathbb{Z}_2[\mathfrak{b}^{(1)}_{a}], \tag{4.20}$$

which are illustrated in Fig. 4.3b.

4.1.3 Compatibility Conditions

In Sect. 3.3.2, we introduced compatibility conditions, that is, constraints on the zero-dimensional topological invariants. In this subsection, we discuss how to formulate compatibility conditions in this terminology.

Before proceeding to the discussion for general situations, we present the compatibility conditions in the one-dimensional even-parity superconductors, as introduced in Sect. 4.1.2. As outlined in Sect. 4.1.2, Pfaffian invariants are assigned to each parity sector at Γ and X. It is important to note that $P_k^+ + P_k^-$ (for $k = \Gamma$, X) represents

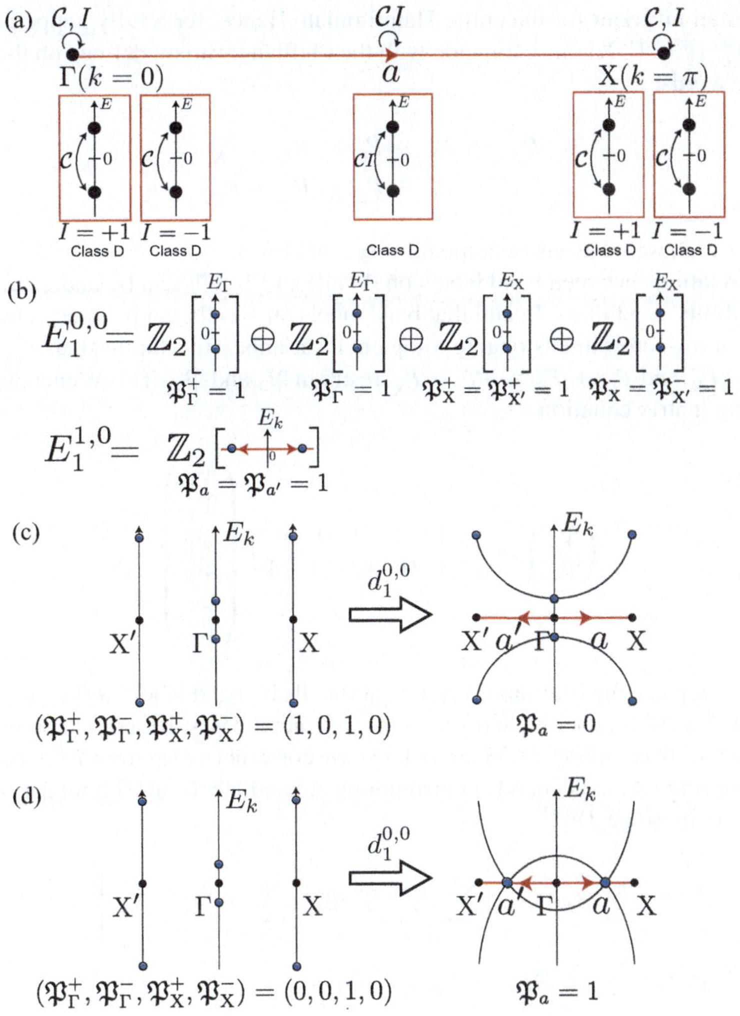

Fig. 4.3 One-dimensional even-parity superconductors. **a** A fundamental domain of BZ and the EAZ classes for cells in $F^0 = \{\Gamma, \mathrm{X}\}$ and $F^1 = \{a\}$, with red arrows indicating the orientations of the 1-cell. **b** An illustrative explanation of $E_1^{0,0}$ and $E_1^{1,0}$, where the bracketed entries denote the band structures of the generators. **c**, **d** An illustration of the physical process of $d_1^{0,0}$. For a system characterized by $(\mathfrak{P}_\Gamma^+, \mathfrak{P}_\Gamma^-, \mathfrak{P}_X^+, \mathfrak{P}_X^-) = (1, 0, 1, 0)$, $d_1^{0,0}$ does not lead to a gapless point on the 1-cells, as shown in (c). Conversely, for a system with $(\mathfrak{P}_\Gamma^+, \mathfrak{P}_\Gamma^-, \mathfrak{P}_X^+, \mathfrak{P}_X^-) = (0, 0, 1, 0)$, a gapless point must appear on each 1-cell, as illustrated in (d). In these figures, $\mathfrak{P}_X'^+$, $\mathfrak{P}_X'^-$, and $\mathfrak{P}_{a'}$ are omitted since $\mathfrak{P}_X'^{\pm} = \mathfrak{P}_X^{\pm}$ and $\mathfrak{P}_{a'} = \mathfrak{P}_a$. Reprinted from Ref. [1] by The Author(s) licensed under CC BY 4.0

the Pfaffian invariant for the entire Hamiltonian. Hence, for a fully gapped system, $P_k^+ + P_k^-$ ($k = \Gamma, \mathrm{X}$) must coincide with the Pfaffian invariant defined on the 1-cell. In other words,

$$P_{k\in a} = P_\Gamma^+ + P_\Gamma^- = P_\mathrm{X}^+ + P_\mathrm{X}^-, \tag{4.21}$$

$$P_{k\in a'} = P_\Gamma^+ + P_\Gamma^- = P_{\mathrm{X}'}^+ + P_{\mathrm{X}'}^-. \tag{4.22}$$

We refer to these relations as compatibility conditions.

The relations between band labels on 0-cells and 1-cells can be understood from compatibility conditions. Recall that band labels on 1-cells can be interpreted as the change of the zero-dimensional topological invariants. This implies that $P_\Gamma^+ + P_\Gamma^- - P_\mathrm{X}^+ - P_\mathrm{X}^-$ and $P_\Gamma^+ + P_\Gamma^- - P_{\mathrm{X}'}^+ - P_{\mathrm{X}'}^-$ result in $\mathfrak{P}_a$ and $\mathfrak{P}_{a'}$. This is encoded in the following matrix equation

$$\begin{pmatrix} \mathfrak{P}_a \\ \mathfrak{P}_{a'} \end{pmatrix} = \begin{pmatrix} 1 & 1 & -1 & -1 & 0 & 0 \\ 1 & 1 & 0 & 0 & -1 & -1 \end{pmatrix} \begin{pmatrix} \mathfrak{P}_\Gamma^+ \\ \mathfrak{P}_\Gamma^- \\ \mathfrak{P}_\mathrm{X}^+ \\ \mathfrak{P}_\mathrm{X}^- \\ \mathfrak{P}_\mathrm{X}'^+ \\ \mathfrak{P}_\mathrm{X}'^- \end{pmatrix}, \tag{4.23}$$

where we replace the Pfaffian invariants at 0-cells by band labels at 0-cells.

Then, we extend the above discussions to general cases. Based on compatibility conditions defined in Eqs. (3.33) and (3.36), we construct a map from $E_1^{p,0}$ to $E_1^{p+1,0}$. The following relations encode contributions of band labels on all boundary p-cells of D^{p+1} to those on D^{p+1}:

$$\mathfrak{P}_{D^{p+1}}^\beta = \sum_i \delta_{D_i^p, D^{p+1}} \left[\sum_\alpha c_{D_i^p, D^{p+1}}^{\alpha\beta} \mathfrak{P}_k^\alpha + \sum_\gamma c_{D_i^p, D^{p+1}}^{\gamma\beta} \mathfrak{N}_k^\gamma \right], \tag{4.24}$$

$$\mathfrak{N}_{D^{p+1}}^\beta = \sum_i \sum_\alpha \delta_{D_i^p, D^{p+1}} c_{D_i^p, D^{p+1}}^{\alpha\beta} \mathfrak{N}_{D_i^p}^\alpha, \tag{4.25}$$

where $\delta_{D_i^p, D^{p+1}}$ represents the contribution of D_i^p to D^{p+1}. Namely, when D^{p+1} is not adjacent to D_i^p, $\delta_{D_i^p, D^{p+1}} = 0$. If the orientation of D_i^p agrees (disagrees) with the orientation of the $(p+1)$-cell D^{p+1} adjacent to D_i^p, $\delta_{D_i^p, D^{p+1}} = 1\ (-1)$. Also, $c_{D_i^p, D^{p+1}}^{\alpha\beta}$ is the same as the coefficient in Eqs. (3.33) and (3.36). While all coefficients of zero-dimensional topological invariants are non-negative in Eqs. (3.33) and (3.36), some coefficients of band labels in Eqs. (4.24) and (4.25) can take negative values. By applying the above relations to all $(p+1)$-cells, we construct a matrix equation among band labels on p- and $(p+1)$-cells.

Rewriting the matrix constructed by Eqs. (4.24) and (4.25) in terms of basis vectors of $E_1^{p,0}$ to $E_1^{p+1,0}$, we obtain a map from $E_1^{p,0}$ to $E_1^{p+1,0}$

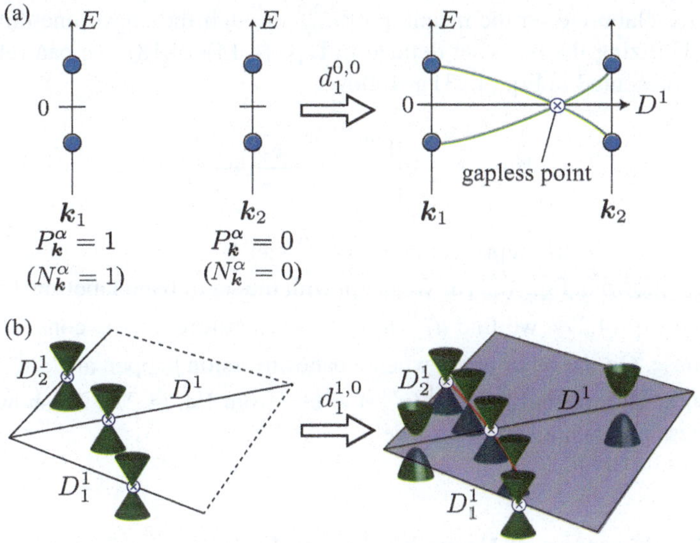

Fig. 4.4 $d_1^{p,0}$. **a** Given a set of zero-dimensional topological invariants at 0-cells, $d_1^{0,0}$ informs us whether gapless points must appear on the adjacent 1-cells or not. The figure focuses on two 0-cells (labeled as $\boldsymbol{k}_1$ and $\boldsymbol{k}_2$) and the 1-cell connecting them. **b** Regarding gapless points on 1-cells, $d_1^{1,0}$ determines if these points should extend to adjacent 2-cells. The figure discusses two 2-cells adjacent to the 1-cell D^1, showing a scenario where the gapless points are extended. Reprinted from Ref. [1] by The Author(s) licensed under CC BY 4.0

$$d_1^{p,0} : E_1^{p,0} \to E_1^{p+1,0}, \tag{4.26}$$

which is dubbed *first differential* [2]. It should be noted that $d_1^{p,0}$ always satisfies $d_1^{p+1,0} \circ d_1^{p,0} = 0$, i.e., $d_1^{p+1,0}\left(d_1^{p,0}(\mathfrak{n}^{(p)})\right) = \mathbf{0}$ for an arbitrary $\mathfrak{n}^{(p)}$.

Since $E_1^{p,0}$ contains only local information about p-cells, it does not determine their global structures. The first differential $d_1^{p,0}$ elucidates the relationship between p and $(p+1)$-cells, as depicted in Fig. 4.4. For $p = 0$, $d_1^{0,0}$ informs us whether gapped states at all 0-cells can be connected without closing the gap on 1-cells or not. When $d_1^{0,0}(\mathfrak{n}^{(0)}) = \mathbf{0}$, all compatibility conditions on zero-dimensional topological invariants at 0-cells hold. Conversely, $d_1^{0,0}(\mathfrak{n}^{(0)}) \neq \mathbf{0}$ implies the violation of some compatibility conditions, indicating the presence of gapless points on the 1-cells. For p greater than one, a nontrivial $d_1^{p,0}$ tells us whether the gapless regions on p-cells are extend to those on $(p+1)$-cells. Specifically, a nontrivial $d_1^{1,0}$ checks whether a gapless point on a 1-cell, corresponding to a generator of $E_1^{1,0}$, extends to the adjacent 2-cells, leading to gapless lines on the 2-cells. Similarly, $d_1^{2,0}$ examine if a gapless line on a 2-cell, corresponding to a generator of $E_1^{2,0}$, is connected to to gapless surfaces on the 3-cells. Sections 4.3.1 and 4.3.2 provide further explanation on utilizing these first differentials to classify gapless points on 1-cells.

Here, we elaborate on the meaning of $d_1^{0,0}$ through the above one-dimensional example. Utilizing the basis set defined in Eqs. (4.14)–(4.18), we can reformulate the matrix presented in Eq. (4.23) as follows:

$$M_{d_1^{0,0}} = \begin{array}{c|cccc} & \mathfrak{b}^{(0)}_{\Gamma,+} & \mathfrak{b}^{(0)}_{\Gamma,-} & \mathfrak{b}^{(0)}_{X,+} & \mathfrak{b}^{(0)}_{X,-} \\ \hline \mathfrak{b}^{(1)}_{a} & 1 & 1 & -1 & -1 \end{array}, \tag{4.27}$$

which serves as a matrix representation of $d_1^{0,0}$. Let us consider two concrete sets of band labels shown in Fig. 4.3c, d. We begin with the set of band labels $\mathfrak{n}^{(0)} = \mathfrak{b}^{(0)}_{\Gamma,+} + \mathfrak{b}^{(0)}_{X,+}$. From Eq. (4.27), we find $d_1^{0,0}(\mathfrak{n}^{(0)}) = 0$. Therefore, this is consistent with a gapped phase. On the other hand, when a band-inversion happen at Γ, $\mathfrak{n}'^{(0)} = \mathfrak{b}^{(0)}_{X,+}$. Correspondingly, we find that $d_1^{0,0}(\mathfrak{n}'^{(0)}) = \mathfrak{b}^{(1)}_{a}$ from Eq. (4.27). As shown in Fig. 4.3d, gapless points must exist on 1-cells.

4.2 Classification of Point Nodes on 1-Cells

As mentioned in the previous section, gapless points on 1-cells are not always genuine point nodes. In this section, we discuss a classification scheme for locally stable point nodes on 1-cells. The effective Hamiltonian around a gapless point on a 1-cell is given by

$$H_{(k_1,k_2)} = k_1\gamma_1 + k_2\gamma_2 + \delta k_3\gamma_0, \tag{4.28}$$

$$\gamma_i\gamma_j + \gamma_j\gamma_i = 2\delta_{ij}\mathbb{1}, \tag{4.29}$$

where k_1 and k_2 are displacements from the gapless point in the directions perpendicular to the 1-cell, δk_3 is that in the direction parallel to the 1-cell. Here, we consider $k_{i=1,2}$ as a momentum parameter and δk_3 as a mass parameter. The energy dispersion of the above Hamiltonian is given by $E_{(k_1,k_2)} = \sqrt{(k_1)^2 + (k_2)^2 + (\delta k_3)^2}$, and thus the Hamiltonian is gapped except for $(k_1, k_2, \delta k_3) = (0, 0, 0)$. The classification of the point nodes on 1-cells of three-dimensional systems is equivalent to that of the above Dirac Hamiltonian.

Reference [15] has shown that one can redefine any point group symmetries as onsite symmetries in such a way that classifications of massive Dirac Hamiltonians do not change. In this thesis, we will refer to it as Cornfeld-Chapman's method. Based on this method, Refs. [15, 16] have classified three-dimensional massive Dirac Hamiltonians in the presence of nonmagnetic and magnetic point group symmetries.

Applying Cornfeld-Chapman's method [15] to the above two-dimensional massive Dirac Hamiltonians, we reveal whether point nodes on 1-cells are stable or not. As a result, we find the following three types of results: (i) The classification is nontrivial, and a gapless point on the line is a point node. (ii) The classification is nontrivial, and a point node is composed of multiple gapless points. (iii) The classification is trivial; thus, a gapless point on the line is part of line or surface nodes. We will integrate the point-node classifications into compatibility conditions discussed in Sect. 4.3.

4.2.1 Cornfeld-Chapman's Method for Two-Dimensional Systems

Consider a gapless point on a 1-cell D^1. We analyze the massive Dirac Hamiltonian near D^1 in Eq. (4.28) with symmetries in the little co-group G_{D^1}/T. The Dirac Hamiltonian $H_{(k_1,k_2)}$ satisfies

$$\sigma_{\mathbf{k}}(g)H_{(k_1,k_2)} = \begin{cases} H_{r_g(k_1,k_2)}\sigma_{\mathbf{k}}(g) & \text{for } c_g = +1, \\ -H_{r_g(k_1,k_2)}\sigma_{\mathbf{k}}(g) & \text{for } c_g = -1, \end{cases} \tag{4.30}$$

where r_g is an element of two-dimensional orthogonal group O(2). Generally speaking, r_g is written by

$$r_g = \begin{cases} \begin{pmatrix} \cos\theta_g & -\sin\theta_g \\ \sin\theta_g & \cos\theta_g \end{pmatrix} & \text{for } \det r_g = +1, \\ \begin{pmatrix} -\cos\theta_g & -\sin\theta_g \\ -\sin\theta_g & \cos\theta_g \end{pmatrix} & \text{for } \det r_g = -1. \end{cases} \tag{4.31}$$

Hereafter, $\det r_g$ is denoted by s_g for simplicity.

In Cornfeld-Chapman's method, we can construct onsite symmetries corresponding to the original point group symmetries. Here, we introduce onsite symmetries whose representations are defined by

$$\widetilde{\sigma}(g) \equiv \gamma_1^{\frac{1-s_g}{2}} e^{\frac{\theta_g}{2}\gamma_1\gamma_2}\sigma_{\mathbf{k}}(g) \quad \text{for } \forall g \in G_{D^1}/T. \tag{4.32}$$

One can see that

$$\widetilde{\sigma}(g)H_{(k_1,k_2)} = s_g c_g H_{(k_1,k_2)}\widetilde{\sigma}(g), \tag{4.33}$$

$$\widetilde{\sigma}(g)\widetilde{\sigma}(h) = (s_g c_g)^{\frac{1-s_h}{2}} z'_{g,h} z^{\mathbf{k}}_{g,h}\widetilde{\sigma}(gh), \tag{4.34}$$

where $z'_{g,h}$ is determined by

$$\gamma_1^{\frac{1-s_h}{2}} e^{\frac{\theta_h}{2}\gamma_1\gamma_2}\gamma_1^{\frac{1-s_g}{2}} e^{\frac{\theta_g}{2}\gamma_1\gamma_2} = z'_{g,h}\gamma_1^{\frac{1-s_{gh}}{2}} e^{\frac{\theta_{gh}}{2}\gamma_1\gamma_2}. \tag{4.35}$$

It should be noted that the commutation relations between $\widetilde{\sigma}(g)$ and $H_{(k_1,k_2)}$ are different from those between $\sigma_{\boldsymbol{k}}(g)$ and $H_{(k_1,k_2)}$. When $\sigma_{\boldsymbol{k}}(g)$ with $s_g = -1$ commutes (anticommutes) with $H_{(k_1,k_2)}$, $\widetilde{\sigma}(g)$ anticommutes (commutes) with $H_{(k_1,k_2)}$. Similarly, commutation relations between antiunitary symmetries and $H_{(k_1,k_2)}$ also change. As a result, the decomposition of the little co-group is also changed as $G_{D^1}/T = \widetilde{\mathcal{G}} + \widetilde{\mathcal{A}} + \widetilde{\mathcal{P}} + \widetilde{\mathcal{J}}$, where each subset is defined by

$$\widetilde{\mathcal{G}} = \{g \in G_{D^1}/T | s_g c_g = 1, \phi_g = 1\}, \tag{4.36}$$

$$\widetilde{\mathcal{A}} = \{g \in G_{D^1}/T | s_g c_g = 1, \phi_g = -1\}, \tag{4.37}$$

$$\widetilde{\mathcal{P}} = \{g \in G_{D^1}/T | s_g c_g = -1, \phi_g = -1\}, \tag{4.38}$$

$$\widetilde{\mathcal{J}} = \{g \in G_{D^1}/T | s_g c_g = -1, \phi_g = 1\} \tag{4.39}$$

As discussed in Ref. [17], the classification of two-dimensional Dirac Hamiltonians with onsite symmetries is achieved by the second homotopy group of the classifying space. Thus, we have to identify the classifying space. As is the case with Eqs. (2.29) and (2.30), we can simultaneously block-diagonalize $\widetilde{\sigma}(g)$ $(g \in \widetilde{\mathcal{G}})$ and $H_{(k_1,k_2)}$ in such a way that

$$\widetilde{\sigma}(g) = \mathrm{diag}\left[\tilde{u}^{\widetilde{\alpha}_1}(g) \otimes \mathbb{1}_{m_1}, \dots, \tilde{u}^{\widetilde{\alpha}_n}(g) \otimes \mathbb{1}_{m_n}\right], \tag{4.40}$$

$$H_{(k_1,k_2)} = \mathrm{diag}\left[\mathbb{1}_{d^{\widetilde{\alpha}_1}} \otimes h^{\widetilde{\alpha}_1}, \dots, \mathbb{1}_{d^{\widetilde{\alpha}_n}} \otimes h^{\widetilde{\alpha}_n}\right], \tag{4.41}$$

where $\tilde{u}^{\widetilde{\alpha}}(g)$ is an irreducible representations of $\widetilde{\mathcal{G}}$. Here, $d^{\widetilde{\alpha}}$ and $m_{\widetilde{\alpha}}$ are dimensions of $\tilde{u}^{\widetilde{\alpha}}(g)$ and $h^{\widetilde{\alpha}}_{\boldsymbol{k}}$, respectively.

Classifying spaces depend on EAZ classes of each irreducible representation of $\widetilde{\mathcal{G}}$. To identify EAZ classes, we again apply the following Wigner criteria to each sector $h^{\widetilde{\alpha}}$:

$$\widetilde{W}^{\widetilde{\alpha}}(\widetilde{\mathcal{P}}) = \frac{1}{|\widetilde{\mathcal{P}}|}\sum_{c \in \widetilde{\mathcal{P}}} (s_c c_c)^{\frac{1-s_c}{2}} z'_{c,c} z^{\boldsymbol{k}}_{c,c} \widetilde{\chi}^{\widetilde{\alpha}}(c^2) \in \{0, \pm 1\}, \tag{4.42}$$

$$\widetilde{W}^{\widetilde{\alpha}}(\widetilde{\mathcal{A}}) = \frac{1}{|\widetilde{\mathcal{A}}|}\sum_{c \in \widetilde{\mathcal{A}}} (s_a c_a)^{\frac{1-s_a}{2}} z'_{a,a} z^{\boldsymbol{k}}_{a,a} \widetilde{\chi}^{\widetilde{\alpha}}(a^2) \in \{0, \pm 1\}, \tag{4.43}$$

$$\widetilde{W}^{\widetilde{\alpha}}(\widetilde{\mathcal{J}}) = \frac{1}{|\widetilde{\mathcal{G}}|}\sum_{g \in \widetilde{\mathcal{G}}} \frac{(s_\gamma c_\gamma)^{\frac{1-s_{\gamma^{-1}g\gamma}}{2}} z'_{\gamma,\gamma^{-1}g\gamma} z^{\boldsymbol{k}}_{\gamma,\gamma^{-1}g\gamma}}{(s_g c_g)^{\frac{1-s_\gamma}{2}} z'_{g,\gamma} z^{\boldsymbol{k}}_{g,\gamma}} [\widetilde{\chi}^{\widetilde{\alpha}}(\gamma^{-1}g\gamma)]^* \widetilde{\chi}^{\widetilde{\alpha}}(g) \in \{0, 1\}, \tag{4.44}$$

where we replace $z^{\boldsymbol{k}}_{g,h}$ in Eqs. (3.23)–(3.25) with $(s_g c_g)^{\frac{1-s_h}{2}} z'_{g,h} z^{\boldsymbol{k}}_{g,h}$. Here, $\widetilde{\chi}^{\widetilde{\alpha}}(g) = \mathrm{tr}[\widetilde{u}^{\widetilde{\alpha}}(g)]$ and γ is an element of $\widetilde{\mathcal{J}}$. Correspondences between EAZ classes and classifying spaces (denoted by C_s and R_s) are summarized in Table 4.1. The classification of two-dimensional Dirac Hamiltonians in Eq. (4.28) is given by the direct sum of $\pi_2(\mathrm{C}_s)$ and $\pi_2(\mathrm{R}_s)$ for each irreducible representation $\tilde{u}^{\widetilde{\alpha}}$ [17].

Table 4.1 EAZ symmetry classes, their classifying spaces, and the second homotopy π_2

EAZ	$\widetilde{\mathcal{W}}[\tilde{\alpha}]$	classifying space	π_2
A, $A_{\mathcal{T}}$, $A_{\mathcal{C}}$, A_{Γ}, $A_{\mathcal{T},\mathcal{C}}$	(0, 0, 0)	C_0	$\mathbb{Z}$
AIII, $AIII_{\mathcal{T}}$	(0, 0, 1)	C_1	0
AI, $AI_{\mathcal{C}}$	(1, 0, 0)	R_0	$\mathbb{Z}_2$
BDI	(1, 1, 1)	R_1	0
D, $D_{\mathcal{T}}$	(0, 1, 0)	R_2	$2\mathbb{Z}$
DIII	(−1, 1, 1)	R_3	0
AII, $AII_{\mathcal{C}}$	(−1, 0, 0)	R_4	0
CII	(−1, −1, 1)	R_5	0
C, $C_{\mathcal{T}}$	(0, −1, 0)	R_6	$\mathbb{Z}$
CI	(1, −1, 1)	R_7	$\mathbb{Z}_2$

The subscripts $\mathcal{T}$, $\mathcal{C}$, and Γ denote that an irreducible representation is related to another one by the onsite antiunitary and/or the chiral symmetries. Reprinted from Ref. [1] by The Author(s) licensed under CC BY 4.0

4.2.2 Character Decomposition Formulas

As discussed in the preceding subsection, we can classify the Dirac Hamiltonians near gapless points on 1-cells. The next step is to construct a map from the generating Dirac Hamiltonians to elements of $E_1^{1,0}$. In this subsection, we derive formulas to obtain elements of $E_1^{1,0}$ from generating Dirac Hamiltonians. The formulas are summarized in Table 4.2.

Consider we have a generating Dirac Hamiltonian on a 1-cell and onsite symmetries in Eq. (4.32). From the definition of these onsite symmetries, we can find the original symmetry representations of G_{D^1}/T as

$$\sigma_k(g) = e^{-\frac{\theta_g}{2}\gamma_1\gamma_2}\gamma_1^{\frac{1-s_g}{2}}\tilde{\sigma}(g). \tag{4.45}$$

Then, we can read how many irreducible representations contained in the above representation $\sigma_k(g)$ in Eq. (4.45), which result in band labels on the 1-cell. From the orthogonality of irreducible representations, $\mathfrak{N}^{\alpha}_{D^1}$ and $\mathfrak{P}^{\alpha}_{D^1}$ are determined by

$$\mathfrak{N}^{\beta}_{D^1} = \frac{1}{|\mathcal{G}_k/T|}\sum_{g\in\mathcal{G}_k/T}\chi^{\beta}_k(g)\mathrm{tr}[\gamma_0 e^{-\frac{\theta_g}{2}\gamma_1\gamma_2}\gamma_1^{\frac{1-s_g}{2}}\tilde{\sigma}(g)], \tag{4.46}$$

$$\mathfrak{P}^{\beta}_{D^1} = \frac{1/2}{|\mathcal{G}_k/T|}\sum_{g\in\mathcal{G}_k/T}\chi^{\beta}_k(g)\mathrm{tr}[e^{-\frac{\theta_g}{2}\gamma_1\gamma_2}\gamma_1^{\frac{1-s_g}{2}}\tilde{\sigma}(g)] \mod 2. \tag{4.47}$$

Thanks to the insertion of γ_0, occupied and unoccupied bands contribute to $\mathbb{Z}$-valued band labels with opposite signs by γ_0 in Eq. (4.46). After applying this procedure to all irreducible representations of $\mathcal{G}_k$, we map the generating Dirac Hamiltonian to an element of $E_1^{1,0}$.

Table 4.2 Formulas to obtain elements of $E_1^{1,0}$ corresponding to generating Dirac Hamiltonians

EAZ	Formula for the map to $\mathbb{Z}$	Formula for the map to $\mathbb{Z}_2$
A	$\frac{-2i}{\lvert G_{D^1}/T\rvert}\sum_{g\in G_{D^1}}\delta_{s_g,1}\sin\frac{\theta_g}{2}[\chi^{\beta}_{D^1}(g)]^*\widetilde{\chi}^{\widetilde{\alpha}}(g)$	$\frac{1}{\lvert G_{D^1}/T\rvert}\sum_{g\in G_{D^1}}\delta_{s_g,1}\cos\frac{\theta_g}{2}[\chi^{\beta}_{D^1}(g)]^*\widetilde{\chi}^{\widetilde{\alpha}}(g)$
$\mathrm{A}_{\mathcal{T}}$	$\frac{-2i}{\lvert G_{D^1}/T\rvert}\sum_{g\in G_{D^1}}\delta_{s_g,1}\sin\frac{\theta_g}{2}[\chi^{\beta}_{D^1}(g)]^*\left(\widetilde{\chi}^{\widetilde{\alpha}}(g)-\widetilde{\chi}^{\mathcal{T}\widetilde{\alpha}}(g)\right)$	$\frac{1}{\lvert G_{D^1}/T\rvert}\sum_{g\in G_{D^1}}\delta_{s_g,1}\cos\frac{\theta_g}{2}[\chi^{\beta}_{D^1}(g)]^*\left(\widetilde{\chi}^{\widetilde{\alpha}}(g)+\widetilde{\chi}^{\mathcal{T}\widetilde{\alpha}}(g)\right)$
$\mathrm{A}_{\mathcal{C}}$	$\frac{-2i}{\lvert G_{D^1}/T\rvert}\sum_{g\in G_{D^1}}\delta_{s_g,1}\sin\frac{\theta_g}{2}[\chi^{\beta}_{D^1}(g)]^*\left(\widetilde{\chi}^{\widetilde{\alpha}}(g)+\widetilde{\chi}^{\mathcal{C}\widetilde{\alpha}}(g)\right)$	$\frac{1}{\lvert G_{D^1}/T\rvert}\sum_{g\in G_{D^1}}\delta_{s_g,1}\cos\frac{\theta_g}{2}[\chi^{\beta}_{D^1}(g)]^*\left(\widetilde{\chi}^{\widetilde{\alpha}}(g)+\widetilde{\chi}^{\mathcal{C}\widetilde{\alpha}}(g)\right)$
A_{Γ}	$\frac{-2i}{\lvert G_{D^1}/T\rvert}\sum_{g\in G_{D^1}}\delta_{s_g,1}\sin\frac{\theta_g}{2}[\chi^{\beta}_{D^1}(g)]^*\left(\widetilde{\chi}^{\widetilde{\alpha}}(g)-\widetilde{\chi}^{\Gamma\widetilde{\alpha}}(g)\right)$	$\frac{1}{\lvert G_{D^1}/T\rvert}\sum_{g\in G_{D^1}}\delta_{s_g,1}\cos\frac{\theta_g}{2}[\chi^{\beta}_{D^1}(g)]^*\left(\widetilde{\chi}^{\widetilde{\alpha}}(g)+\widetilde{\chi}^{\Gamma\widetilde{\alpha}}(g)\right)$
$\mathrm{A}_{\mathcal{T},\mathcal{C}}$	$\frac{-2i}{\lvert G_{D^1}/T\rvert}\sum_{g\in G_{D^1}}\delta_{s_g,1}\sin\frac{\theta_g}{2}[\chi^{\beta}_{D^1}(g)]^*$ $\times\left(\widetilde{\chi}^{\widetilde{\alpha}}(g)-\widetilde{\chi}^{\mathcal{T}\widetilde{\alpha}}(g)+\widetilde{\chi}^{\mathcal{C}\widetilde{\alpha}}(g)-\widetilde{\chi}^{\Gamma\widetilde{\alpha}}(g)\right)$	$\frac{1}{\lvert G_{D^1}/T\rvert}\sum_{g\in G_{D^1}}\delta_{s_g,1}\cos\frac{\theta_g}{2}[\chi^{\beta}_{D^1}(g)]^*$ $\times\left(\widetilde{\chi}^{\widetilde{\alpha}}(g)+\widetilde{\chi}^{\mathcal{T}\widetilde{\alpha}}(g)+\widetilde{\chi}^{\mathcal{C}\widetilde{\alpha}}(g)+\widetilde{\chi}^{\Gamma\widetilde{\alpha}}(g)\right)$
C	$\frac{-2i}{\lvert G_{D^1}/T\rvert}\sum_{g\in G_{D^1}}\delta_{s_g,1}\sin\frac{\theta_g}{2}[\chi^{\beta}_{D^1}(g)]^*\widetilde{\chi}^{\widetilde{\alpha}}(g)$	$\frac{1}{\lvert G_{D^1}/T\rvert}\sum_{g\in G_{D^1}}\delta_{s_g,1}\cos\frac{\theta_g}{2}[\chi^{\beta}_{D^1}(g)]^*\widetilde{\chi}^{\widetilde{\alpha}}(g)$
$\mathrm{C}_{\mathcal{T}}$	$\frac{-2i}{\lvert G_{D^1}/T\rvert}\sum_{g\in G_{D^1}}\delta_{s_g,1}\sin\frac{\theta_g}{2}[\chi^{\beta}_{D^1}(g)]^*\left(\widetilde{\chi}^{\widetilde{\alpha}}(g)-\widetilde{\chi}^{\mathcal{T}\widetilde{\alpha}}(g)\right)$	$\frac{1}{\lvert G_{D^1}/T\rvert}\sum_{g\in G_{D^1}}\delta_{s_g,1}\cos\frac{\theta_g}{2}[\chi^{\beta}_{D^1}(g)]^*\left(\widetilde{\chi}^{\widetilde{\alpha}}(g)+\widetilde{\chi}^{\mathcal{T}\widetilde{\alpha}}(g)\right)$
D	$\frac{-4i}{\lvert G_{D^1}/T\rvert}\sum_{g\in G_{D^1}}\delta_{s_g,1}\sin\frac{\theta_g}{2}[\chi^{\beta}_{D^1}(g)]^*\widetilde{\chi}^{\widetilde{\alpha}}(g)$	$\frac{2}{\lvert G_{D^1}/T\rvert}\sum_{g\in G_{D^1}}\delta_{s_g,1}\cos\frac{\theta_g}{2}[\chi^{\beta}_{D^1}(g)]^*\widetilde{\chi}^{\widetilde{\alpha}}(g)$
$\mathrm{D}_{\mathcal{T}}$	$\frac{4i}{\lvert G_{D^1}/T\rvert}\sum_{g\in G_{D^1}}\delta_{s_g,1}\sin\frac{\theta_g}{2}[\chi^{\beta}_{D^1}(g)]^*\left(\widetilde{\chi}^{\widetilde{\alpha}}(g)-\widetilde{\chi}^{\mathcal{T}\widetilde{\alpha}}(g)\right)$	$\frac{2}{\lvert G_{D^1}/T\rvert}\sum_{g\in G_{D^1}}\delta_{s_g,1}\cos\frac{\theta_g}{2}[\chi^{\beta}_{D^1}(g)]^*\left(\widetilde{\chi}^{\widetilde{\alpha}}(g)+\widetilde{\chi}^{\mathcal{T}\widetilde{\alpha}}(g)\right)$
AI	–	$\frac{2}{\lvert G_{D^1}/T\rvert}\sum_{g\in G_{D^1}}\delta_{s_g,1}\cos\frac{\theta_g}{2}[\chi^{\beta}_{D^1}(g)]^*\widetilde{\chi}^{\widetilde{\alpha}}(g)$
$\mathrm{AI}_{\mathcal{C}}$	–	$\frac{2}{\lvert G_{D^1}/T\rvert}\sum_{g\in G_{D^1}}\delta_{s_g,1}\cos\frac{\theta_g}{2}[\chi^{\beta}_{D^1}(g)]^*\left(\widetilde{\chi}^{\widetilde{\alpha}}(g)+\widetilde{\chi}^{\mathcal{C}\widetilde{\alpha}}(g)\right)$
CI	–	$\frac{2}{\lvert G_{D^1}/T\rvert}\sum_{g\in G_{D^1}}\delta_{s_g,1}\cos\frac{\theta_g}{2}[\chi^{\beta}_{D^1}(g)]^*\widetilde{\chi}^{\widetilde{\alpha}}(g)$

In this table, χ^{α} and $\widetilde{\chi}^{\widetilde{\alpha}}$ represent the characters of the little co-group $\mathcal{G}_{D^1}/T$ and the onsite symmetry group $\widetilde{\mathcal{G}}$, respectively. The first column indicates the EAZ classes of the irreducible representations denoted by $\widetilde{u}^{\widetilde{\alpha}}$. Furthermore, $\mathcal{T}\widetilde{\alpha}$, $\mathcal{C}\widetilde{\alpha}$, and $\Gamma\widetilde{\alpha}$ label the irreducible representations related to $\widetilde{\alpha}$ by onsite time-reversal, particle-hole, and chiral symmetries, respectively. The derivations of these formulas can be found in Appendix 2. Reprinted from Ref. [1] by The Author(s) licensed under CC BY 4.0

For each of EAZ classes, we can fix the form of generating Hamiltonians and representations. Then, we have derived the formulas summarized in Table 4.2. For instance, we here derive the formulas for class $\mathrm{A}_{\mathcal{C}}$. The generating Hamiltonian and representations are denoted by

$$H_{(k_1,k_2)} = k_1\tau_1 + k_2\tau_2 + \delta k_3\tau_3, \tag{4.48}$$

$$\widetilde{\sigma}(\mathcal{C}) = \begin{cases} i\tau_2\sigma_1 K & \text{for } [\widetilde{\sigma}(\mathcal{C})]^2 = -1, \\ \tau_2\sigma_2 K & \text{for } [\widetilde{\sigma}(\mathcal{C})]^2 = +1, \end{cases} \tag{4.49}$$

$$\widetilde{\sigma}(g) = \tau_0 \begin{pmatrix} \widetilde{u^{\tilde{\alpha}}}(g) & \\ & \widetilde{u^{\widetilde{\mathcal{C}\alpha}}}(g) \end{pmatrix} \quad \text{for } \forall \widetilde{g} \in \widetilde{\mathcal{G}}, \tag{4.50}$$

where $\widetilde{u}^{\widetilde{\mathcal{C}\alpha}}$ is an irreducible representation related to $\widetilde{u}^{\tilde{\alpha}}$ by onsite particle-hole symmetries in $\widetilde{\mathcal{P}}$. Also, $\widetilde{\sigma}(\mathcal{C})$ is the generator of $\widetilde{\mathcal{P}}$, and σ_μ and τ_μ ($\mu = 0, 1, 2, 3$) are Pauli matrices originating from different degrees of freedom. By applying the formulas in Eqs. (4.46) and (4.47) to $\gamma_0 = \tau_3$ and representations in Eq. (4.50), we have

$$\mathfrak{N}^{\beta}_{D^1} = \frac{-2i}{|G_{D^1}/T|} \sum_{g\in\mathcal{G}_{D^1}/T} \delta_{s_g,1} \sin\frac{\theta_g}{2}[\chi^{\beta}_{D^1}(g)]^* \left(\widetilde{\chi}^{\tilde{\alpha}}(g) + \widetilde{\chi}^{\widetilde{\mathcal{C}\alpha}}(g)\right), \tag{4.51}$$

$$\mathfrak{P}^{\beta}_{D^1} = \frac{1}{|G_{D^1}/T|} \sum_{g\in\mathcal{G}_{D^1}/T} \delta_{s_g,1} \cos\frac{\theta_g}{2}[\chi^{\beta}_{D^1}(g)]^* \left(\widetilde{\chi}^{\tilde{\alpha}}(g) + \widetilde{\chi}^{\widetilde{\mathcal{C}\alpha}}(g)\right) \bmod 2, \tag{4.52}$$

where $\widetilde{\chi}^{\tilde{\alpha}}(g) = \mathrm{tr}[\widetilde{u}^{\tilde{\alpha}}(g)]$.

Finally, we arrive at three types of results:

(i) The classification of Dirac Hamiltonians is nontrivial, and a generating Dirac Hamiltonian is mapped to a generator of $E_1^{1,0}$. In this case, the gapless point corresponding to the generator of $E_1^{1,0}$ is a genuine point node.
(ii) The classification of Dirac Hamiltonians is nontrivial, but the mapped element of $E_1^{1,0}$ does not coincide with any generator of $E_1^{1,0}$. In other words, the obtained element is a linear combination of multiple generators of $E_1^{1,0}$, and the realized point node must be composed of gapless points on the 1-cell. However, these gapless points are not forced to be at the same momentum by symmetries. In such a case, point nodes are shrinkable loop or surface nodes.
(iii) The classification of Dirac Hamiltonians is trivial. The trivial classification indicates that there are no point and shrinkable nodes.

One might notice that the degeneracy of a point node does not coincide with the dimension of the corresponding Dirac Hamiltonian for Case (i). This difference originates from the fact that we restrict ourselves to linearly dispersive Dirac Hamiltonians. In fact, some states are gappable with symmetric perturbations. Then, the degeneracy of a point node is the same as that of the gapless point in Dirac Hamiltonians with the perturbations. The existence of linearly dispersive Dirac Hamiltonians in Eq. (4.28) ensures that the point node is stable against adding trivial degrees of freedom, which is stability in the K-theoretic sense. One might think that our scheme has missed nodes such that they are line or surface nodes when no trivial degrees of freedom are added. This stability is in the spirit of fragile topological phases [3]. However, quadratic and cubic terms enable us to construct Dirac Hamiltonians with minimal dimensions. Therefore, we conclude that such fragile nodes do not exist on 1-cells.

4.2.3 Example

It is instructive to discuss examples. In this subsection, we consider four symmetry settings of spinful electrons: magnetic space groups $P2/m1'$ with B_g pairing, $P21'$ with B pairing, and $P4$ with ${}^1E$ pairing. Following the discussions in Sects. 4.2.1 and 4.2.2, we obtain classifications of the Dirac Hamiltonians in Eq. (4.28) and elements of $E_1^{1,0}$ for generating Dirac Hamiltonians. The presented results in this subsection will be used to classify gapless points on 1-cells in Sect. 4.3.3.

$P2/m1'$ with B_g Pairing

Our first example is magnetic space group $P2/m1'$ with B_g pairing. Recall that this magnetic space group has twofold rotation C_2^y along the y-axis, inversion I, and TRS $\mathcal{T}$. For this pairing symmetry, PHS anticommutes with twofold rotation but commutes with inversion; that is $\sigma(\mathcal{C})[\sigma(C_2^y)]^* + \sigma(C_2^y)\sigma(\mathcal{C}) = 0$ and $\sigma(\mathcal{C})[\sigma(I)]^* - \sigma(I)\sigma(\mathcal{C}) = 0$ hold. Let us consider a rotation-symmetric line as 1-cell D^1 shown in Fig. 4.5a. Its little co-group is decomposed into the following subsets:

$$\mathcal{G}_{D^1}/T = \{e, C_2^y\} \quad (e : \text{identity}), \tag{4.53}$$

$$\mathcal{A}_{D^1}/T = \{I\mathcal{T}, (IC_2^y)\mathcal{T}\}, \tag{4.54}$$

$$\mathcal{P}_{D^1}/T = \{I\mathcal{C}, (IC_2^y)\mathcal{C}\}, \tag{4.55}$$

$$\mathcal{J}_{D^1}/T = \{\Gamma \equiv \mathcal{T}\mathcal{C}, C_2^y\Gamma\}. \tag{4.56}$$

To carry out the procedures discussed in Sect. 4.2.1, we define generators of the onsite symmetry group by

$$\tilde{\sigma}(C_2^y) \equiv \gamma_1\gamma_2\sigma(C_2^y), \tag{4.57}$$

$$\tilde{\sigma}(I\mathcal{T}) \equiv \sigma(I\mathcal{T}), \tag{4.58}$$

$$\tilde{\sigma}(I\mathcal{C}) \equiv \sigma(I\mathcal{C}), \tag{4.59}$$

One can see that $s_g = +1$ for e and C_2^y, and thus the onsite unitary symmetry group is $\tilde{\mathcal{G}} = \{e, C_2^y\}$. Since $[\tilde{\sigma}(C_2^y)]^2 = -[\sigma(C_2^y)]^2 = +1$, there exist two one-dimensional irreducible representations denoted by $\tilde{u}^{\tilde{\alpha}}(C_2^y) = \alpha$ ($\alpha = \pm 1$). Commutation and anticommutation relations among the representations in Eqs. (4.57)–(4.59) are the same as those among $\sigma(C_2^y)$, $\sigma(I\mathcal{T})$, and $\sigma(I\mathcal{C})$, i.e.,

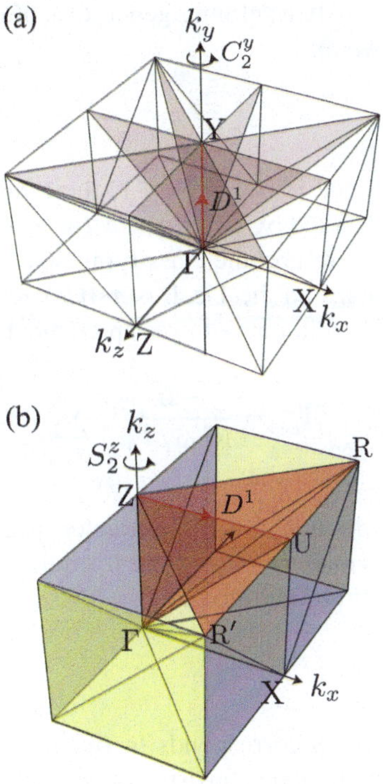

Fig. 4.5 Cell decomposition for the half BZ in $P2/m1'$. Here, we do not show orientations except for the 1-cells denoted by D^1. Adjacent 2-cells to the 1-cell D^1 are colored red. Reprinted from Ref. [1] by The Author(s) licensed under CC BY 4.0

$$\{\tilde{\sigma}(C_2^y), \tilde{\sigma}(IC)\} = 0, \tag{4.60}$$

$$[\tilde{\sigma}(C_2^y), \tilde{\sigma}(IT)] = 0, \tag{4.61}$$

$$[\tilde{\sigma}(IT)]^2. \tag{4.62}$$

From the above equations, we find that EAZ classes for $\alpha = \pm 1$ are class AII_C and that the classifying space is R_4, whose classification is $\pi_2(\mathrm{R}_4) = 0$. The trivial classification is Case (iii); therefore, no point nodes are stable on this line. We show that the gapless point on the line is part of line nodes in Sect. 4.3.

$P21'$ with B Pairing

Next, we discuss magnetic space group $P21'$ with B pairing, whose generators are twofold rotation C_2^y along the y-axis and TRS $\mathcal{T}$. For this pairing symmetry, the relation $\sigma(\mathcal{C})[\sigma(C_2^y)]^* + \sigma(C_2^y)\sigma(\mathcal{C}) = 0$ holds. Again, we consider a rotation-symmetric line as 1-cell D^1 in Fig. 4.5a. Unlike the case of magnetic space group $P2/m1'$, the little co-group on D^1 is decomposed into

$$\mathcal{G}_{D^1}/T = \{e, C_2^y\}, \tag{4.63}$$

$$\mathcal{J}_{D^1}/T = \{\Gamma, C_2^y\Gamma\}. \tag{4.64}$$

After defining generators of onsite symmetries by Eq. (4.57) and $\widetilde{\sigma}(\Gamma) = \sigma(\Gamma)$, we find

$$\{\widetilde{\sigma}(\Gamma), \widetilde{\sigma}(C_2^y)\} = 0. \quad (4.65)$$

Since $[\widetilde{\sigma}(C_2^y)]^2 = +1$, there are two one-dimensional irreducible representations denoted by $\widetilde{u}^{\widetilde{\alpha}}(C_2^y) = \alpha$ $(\alpha = \pm 1)$ whose EAZ classes are class A_Γ [see Table 4.3]. Therefore, the classifying space is the C_0, and the classification of Dirac Hamiltonians on the 1-cell is $\pi_2(C_0) = \mathbb{Z}$. The final step is to identify the generating Dirac Hamiltonian of $\mathbb{Z}$ as an element of $E_1^{1,0}$. This can be accomplished by

$$\mathfrak{N}^{\beta}_{D^1} = \frac{-2i}{|G_{D^1}/T|} \sum_{g \in G_{D^1}/T} \delta_{s_g,1} \sin \frac{\theta_g}{2} [\chi^{\beta}_{D^1}(g)]^* \left(\widetilde{\chi}^{\widetilde{\alpha}}(g) - \widetilde{\chi}^{\Gamma\widetilde{\alpha}}(g) \right), \quad (4.66)$$

where $\widetilde{\chi}^{\Gamma\widetilde{\alpha}}$ is the character of an irreducible representation related to $\widetilde{\chi}^{\widetilde{\alpha}}$ by chiral symmetry. Using the formula of Eq. (4.66), we get band labels of the generating Dirac Hamiltonian

$$(\mathfrak{N}^1, \mathfrak{N}^2) = (2, -2), \quad (4.67)$$

which corresponds to twice of a generator of $E_1^{1,0}$. This result indicates that the point node consists of two gapless points and that the point node becomes a loop or surface node in the presence of symmetric perturbations, which is Case (ii). To see this concretely, we consider the following Dirac Hamiltonian

$$H_{(k_1,k_2)} = k_1\tau_1 + k_2\tau_2\sigma_3 + \delta k_3\tau_3, \quad (4.68)$$

$$\widetilde{\sigma}(C_2) = \sigma_3, \quad (4.69)$$

$$\sigma(C_2) = e^{-i\frac{\pi}{2}\tau_3\sigma_3}\widetilde{\sigma}(C_2) = -i\tau_3, \quad (4.70)$$

$$\sigma(\Gamma) = \widetilde{\sigma}(\Gamma) = \tau_2\sigma_2, \quad (4.71)$$

We perturb the Dirac Hamiltonian in Eq. (4.68) by

$$M = m_0\sigma_1 + m_1\sigma_3 + m_2\tau_3 + m_3\tau_3\sigma_2, \quad (4.72)$$

and we find a loop node shown in Fig. 4.6a.

***P*4 with ^1E Pairing**

Next, we discuss magnetic space group $P4$ and focus on a rotation-symmetric line. This is the same 1-cell D^1 in Fig. 4.5a but the axes are exchanged. The little co-group is $G_{D^1}/T = \mathcal{G}_{D^1}/T = \{e, C_4^z, (C_4^z)^2, (C_4^z)^3\}$. Correspondingly, the onsite symmetry group also has a unitary part generated by

$$\widetilde{\sigma}(C_4^z) \equiv e^{\frac{\pi}{4}\gamma_1\gamma_2}\sigma(C_4^z), \quad (4.73)$$

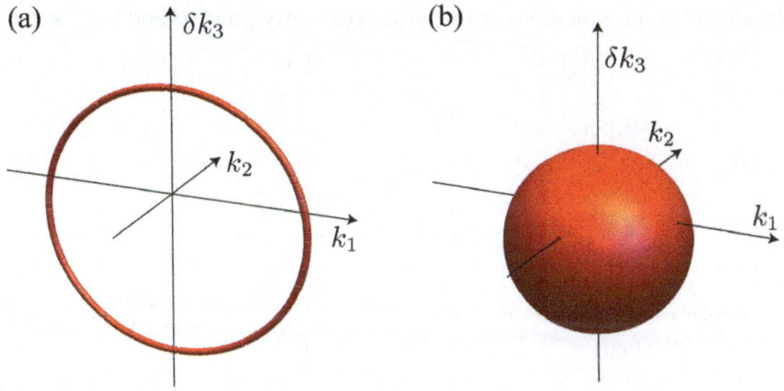

Fig. 4.6 Shrinkable nodes after adding perturbations. **a** shrinkable loop node for Eq. (4.68) in magnetic space group $P21'$ with B pairing, **b** shrinkable surface node for Eq. (4.76) in magnetic space group $P41'$ with ${}^1E$ pairing. Reprinted from Ref. [1] by The Author(s) licensed under CC BY 4.0

Table 4.3 Irreducible representations of $\mathcal{G}_{D^1}/T$ and the onsite unitary group $\widetilde{\mathcal{G}}$ for magnetic space group $P2/m1'$ and $P21'$

	EAZ of $P2/m1'(B_g)$	EAZ of $P21'(B)$	irrep β	e	C_2^y
$\mathcal{G}_{D^1}/T$	D	A	1	1	i
	D	A	2	1	$-i$
	EAZ of $P2/m1'(B_g)$	EAZ of $P21'(B)$	irrep $\widetilde{\alpha}$	e	C_2^y
$\widetilde{\mathcal{G}}$	AII$_C$	A$_\Gamma$	1	1	1
	AII$_C$	A$_\Gamma$	2	1	-1

Reproduced from Table V of Ref. [1]

where $[\widetilde{\sigma}(C_4^z)]^4 = +1$. There are four irreducible representations of $\widetilde{\mathcal{G}}$ shown in Table 4.4. Since their EAZ classes are all class A, the classification of Dirac Hamiltonians on the line is $\mathbb{Z}^4$. For class A, we have the following formula

$$\mathfrak{N}_{D^1}^{\beta} = \frac{-2i}{|G_{D^1}/T|} \sum_{g \in G_{D^1}/T} \delta_{s_g,1} \sin \frac{\theta_g}{2} [\chi_{D^1}^{\beta}(g)]^* \widetilde{\chi}^{\widetilde{\alpha}}(g), \tag{4.74}$$

where β represent to labels of irreducible representations of $\mathcal{G}_{D^1}/T$. Using the formula, we obtain the band labels

$$(\mathfrak{N}_{D^1}^1, \mathfrak{N}_{D^1}^2, \mathfrak{N}_{D^1}^3, \mathfrak{N}_{D^1}^4) = \begin{cases} (-1, 0, 0, 1) & \text{for } \widetilde{\alpha} = 1 \\ (1, -1, 0, 0) & \text{for } \widetilde{\alpha} = 2 \\ (0, 0, 1, -1) & \text{for } \widetilde{\alpha} = 3 \\ (0, 1, -1, 0) & \text{for } \widetilde{\alpha} = 4. \end{cases} \tag{4.75}$$

Table 4.4 Irreducible representations of the onsite symmetry group $\widetilde{\mathcal{G}}$ and $\mathcal{G}_{D^1}/T$ for $P4$

	EAZ	irrep β	e	C_4^z	$(C_4^z)^2$	$(C_4^z)^3$
$\mathcal{G}_{D^1}/T$	A	1	1	$e^{i\frac{\pi}{4}}$	i	$e^{i\frac{3\pi}{4}}$
	A	2	1	$e^{i\frac{3\pi}{4}}$	$-i$	$e^{i\frac{\pi}{4}}$
	A	3	1	$e^{-i\frac{3\pi}{4}}$	i	$e^{-i\frac{\pi}{4}}$
	A	4	1	$e^{-i\frac{\pi}{4}}$	$-i$	$e^{-i\frac{3\pi}{4}}$
	EAZ	irrep $\widetilde{\alpha}$	e	C_4^z	$(C_4^z)^2$	$(C_4^z)^3$
$\widetilde{\mathcal{G}}$	A	1	1	1	1	1
	A	2	1	i	-1	$-i$
	A	3	1	$-i$	-1	i
	A	4	1	-1	1	-1

Reproduced from Table VI of Ref. [1]

They do not correspond to any basis of $E_1^{1,0}$ but coincide with linear combinations of them. Therefore, the result is Case (ii), i.e., the point node is actually a shrunk loop or surface node. Again, let us discuss a concrete Dirac Hamiltonian

$$H_{(k_1,k_2)} = k_1\sigma_1 + k_2\sigma_2 + \delta k_3\sigma_3, \tag{4.76}$$

$$\widetilde{\sigma}(C_4) = \sigma_0, \tag{4.77}$$

$$\sigma(C_4) = e^{-\frac{\pi}{4}\sigma_1\sigma_2}\widetilde{\sigma}(C_4) = \begin{pmatrix} e^{-i\frac{\pi}{4}} & \\ & e^{i\frac{\pi}{4}} \end{pmatrix}, \tag{4.78}$$

which correspond to the case of $\tilde{\alpha} = 1$. $M = \mathrm{diag}(m_0, m_1)$ is a C_4-symmetric perturbation. As shown in Fig. 4.6b, the perturbed Hamiltonian exhibits a shrinkable surface node.

4.3 Unification of Compatibility Conditions and Point-Node Classifications

In this section, by combining classifications of point nodes discussed in Sect. 4.2 with compatibility conditions in Sect. 4.1.3, we provide a unified way to diagnose dimensionalities of nodes. We begin by discussing our classification scheme for gapless points on 1-cells, and then we apply the scheme to several symmetry settings of spinful electrons: magnetic space groups $P2/m1'$ with B_g pairing, $P21'$ with B pairing, and $P4$ with ^1E pairing.

4.3.1 Revisiting Compatibility Conditions and the First Differential

Before moving on to an explanation of our classification scheme, let us revisit the physical meaning of the first differentials $d_1^{p,0}$ for $p = 1, 2$. As discussed in Sect. 4.1.3, nontrivial $d_1^{p,0}$ connects the gapless states on p-cells to those on $(p+1)$-cells. Here, we discuss why they can be interpreted in such a way.

Suppose that we find a gapless point protected by changes of zero-dimensional topological invariants on a 1-cell. In other words, the 1-cell has two parts with different zero-dimensional topological invariants. This does not always mean that the gapless point must be a genuine point node, and it is sometimes part of higher-dimensional nodes. We further consider the case where there exist compatibility conditions between points on the 1-cell and on adjacent 2-cells. Compatibility conditions transform different values of a zero-dimensional topological invariant on the 1-cell into different sets of topological invariants on 2-cells. Then, we have two regions with different zero-dimensional topological invariants on the 2-cells, as shown in Fig. 4.7a. As a result, the boundary of these two regions is a gapless line. The existence or absence of such gapless lines on 2-cells can be understood by $d_1^{1,0}$.

Focusing on one of the adjacent 2-cells, we further apply the same discussion to this 2-cell. In other words, if compatibility conditions between the 2-cell and adjacent 3-cells exist, $d_1^{2,0}$ tells us whether the above two regions are extended to 3-cells or not (see Fig. 4.7b). Consequently, if $d_1^{2,0}$ is nontrivial, the boundary surface is a surface node. In the following, we formulate the above processes in a systematic manner based on $E_1^{p,0}$ and $d_1^{p,0}$.

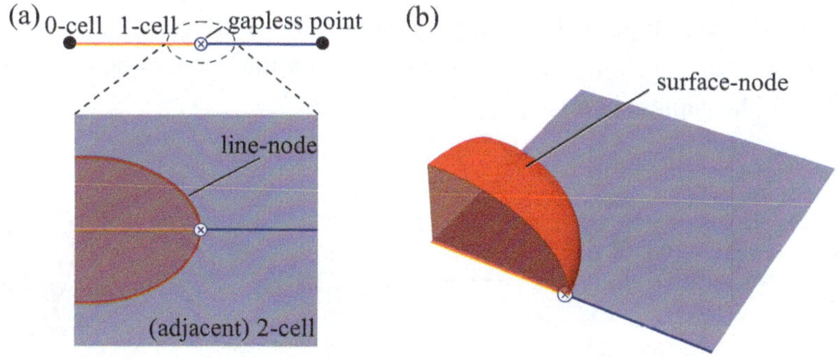

Fig. 4.7 Nodes near a gapless point at a 1-cell. **a** A 1-cell is divided into two parts where topological invariants are different. Compatibility conditions ensure that the region highlighted in red and the surrounding areas exhibit different topological invariant values, leading to a nodal line along their boundary. **b** When compatibility conditions exist between the 2-cell and its adjacent 3-cells, there are two regions whose topological invariants are different in the 3-cells. Consequently, the interface of these regions forms what is known as a surface node. Reprinted from Ref. [1] by The Author(s) licensed under CC BY 4.0

4.3.2 Classifications of Gapless Points on 1-Cells

As discussed above, compatibility conditions inform us of whether the change of zero-dimensional topological invariants on a p-cell induces domain walls of different zero-dimensional topological invariants on $(p+1)$-cells. This process is formulated by generators of $E_1^{p,0}$ and $d_1^{p,0}$. Recall that $E_1^{1,0}$ can be interpreted as the set of gapless states on 1-cells. We start with the set of band labels described by $\mathfrak{n}^{(1)} = \mathfrak{b}^{(1)}_{D^1,\alpha}$, where $\mathfrak{b}^{(1)}_{D^1,\alpha}$ is a basis of $E_1^{1,0}$ generated by an irreducible representation $U^{\alpha}_{D^1}$ on a 1-cell D^1 (see Sect. 4.1.2). Then, we check if compatibility conditions produce gapless lines on 2-cells. This is expressed by $d_1^{1,0}(\mathfrak{n}^{(1)})$. In the following, we separately consider two cases: (A) $d_1^{1,0}(\mathfrak{n}^{(1)}) \neq \mathbf{0}$ and (B) $d_1^{1,0}(\mathfrak{n}^{(1)}) = \mathbf{0}$.

We begin by discussing Case (A). Since $d_1^{1,0}(\mathfrak{n}^{(1)})$ is an element of $E_1^{2,0}$,

$$d_1^{1,0}(\mathfrak{n}^{(1)}) = \sum_i \left(\sum_\alpha r^{(2)}_{D_i^2,\alpha} \mathfrak{b}^{(2)}_{D_i^2,\alpha} + \sum_\beta m^{(2)}_{D_i^2,\beta} \mathfrak{b}^{(2)}_{D_i^2,\beta} \right) \in E_1^{2,0}, \tag{4.79}$$

where $r^{(2)}_{D_i^2,\alpha} \in \mathbb{Z}_2$ and $m^{(2)}_{D_i^2,\beta} \in \mathbb{Z}$. Nonzero coefficients in the generators of $E_1^{2,0}$ in Eq. (4.79) suggest that a gapless point on the 1-cell extends to the corresponding 2-cells, resulting in gapless lines on these 2-cells. Consequently, the gapless point on the 1-cell is identified as part of either line or surface nodes. To determine which one is realized, we further examine whether $d_1^{2,0}$ is nontrivial. One might recall the relation $d_1^{2,0} \circ d_1^{1,0} = 0$ and consider that we cannot use $d_1^{2,0}$ for this purpose. However, this is not the case when focusing on a single adjacent 2-cell. In other words, choosing only one basis, say $\mathfrak{b}^{(2)}_{D_i^2,\alpha}$, from Eq. (4.79), we ask if $d_1^{2,0}\left(\mathfrak{b}^{(2)}_{D_i^2,\alpha}\right) = 0$ or not, as is the case of $E_1^{1,0}$ (see Fig. 4.8 for an intuitive illustration). If there exists at least one basis such that $d_1^{2,0}(\mathfrak{b}^{(2)}_{D_i^2,\alpha}) \neq 0$, the gapless point on the 1-cell is part of a surface node. If not, the gapless point is part of a line node.

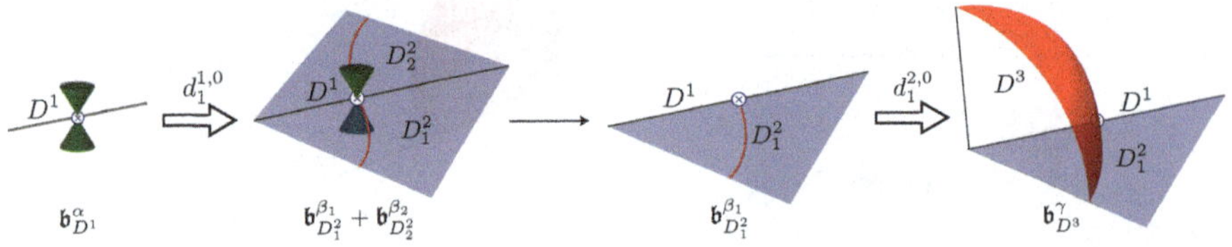

Fig. 4.8 Diagnostic scheme for case (A). First, act $d_1^{1,0}$ on a generator of $E_1^{1,0}$ corresponding to a basis $\mathfrak{b}^{\alpha}_{D^1}$, and then, obtain the gapless lines on the 2-cells adjacent to the 1-cell D^1, which is an element of $E_1^{2,0}$. Second, focus on one of the adjacent 2-cells, say, D_1^2 from the two 2-cells. Correspondingly, we consider only $\mathfrak{b}^{\beta_1}_{D_1^2}$ of $\mathfrak{b}^{\beta_1}_{D_1^2} + \mathfrak{b}^{\beta_2}_{D_2^2}$ in the figure. Finally, we map $\mathfrak{b}^{\beta_1}_{D_1^2}$ to an element $E_1^{3,0}$ by $d_1^{2,0}$ and examine whether the mapped element is trivial or not. Reprinted from Ref. [1] by The Author(s) licensed under CC BY 4.0

Table 4.5 Summary of classification results for examples discussed in this framework

SG	Pairing	HSP1	HSP2	irrep	Classification	Type of node
$P2/m$ with TRS	B_g	$(0, 0, 0)$	$(0, \frac{1}{2}, 0)$	$\bar{\Lambda}_3$	$\mathbb{Z}_2$	L(B)
		$(0, 0, 0)$	$(\frac{1}{2}, 0, 0)$	$\bar{F}_3$	$\mathbb{Z}_2$	L(A)
$P2$ with TRS	B	$(0, 0, 0)$	$(0, \frac{1}{2}, 0)$	$\bar{\Lambda}_3$	$\mathbb{Z}$	L(B)
$P4$ without TRS	${}^1E$	$(0, 0, 0)$	$(0, 0, \frac{1}{2})$	$\bar{\Lambda}_5$	$\mathbb{Z}$	S(A)
				$\bar{\Lambda}_6$	$\mathbb{Z}$	S(A)
				$\bar{\Lambda}_7$	$\mathbb{Z}$	S(A)
				$\bar{\Lambda}_8$	$\mathbb{Z}$	S(A)

The first and second columns in the table present space groups and pairing symmetries, respectively. The third and fourth columns identify the boundary points of 1-cells on which a gapless point is located. The fifth column lists the labels of irreducible representations (irreps), following the notation used in Ref. [4]. The sixth column indicates the classification as either $\mathbb{Z}$ or $\mathbb{Z}_2$, and the seventh column specifies the type of nodes, where 'P', 'L', and 'S' represent point, line, and surface nodes, respectively. Furthermore, the notation (A) is used when the shape of nodes is solely determined by compatibility conditions, whereas (B) indicates that classifications of point nodes are essential. Reproduced from Table VII of Ref. [1]

In Case (B), where $d_1^{1,0}(\mathfrak{n}^{(1)}) = \mathbf{0}$, it is tempting to think that the gapless point on the 1-cell is an actual point node since there are no domain walls discussed above. However, this is not always true. The gapless point on the 1-cell is a genuine point node only if $\mathfrak{n}^{(1)}$ aligns with the classifications of Dirac Hamiltonians, meaning that $\mathfrak{n}^{(1)}$ can be expanded using band labels from point-node classifications as discussed in Sect. 4.2.2. If not, the gapless point on the 1-cell is part of line nodes that extend from the 1-cell to 3-cells.

Ref. [1] performed an exhaustive classification based on the above scheme, and the results are tabulated in the Supplementary Materials of Ref. [1].

4.3.3 *Examples*

In this section, applying the approach explained above to several symmetry classes, we demonstrate the power of our scheme. Based on our method, we first reproduce the results for spinful superconductors in magnetic space group $P2/m1'$ with B_g pairing, which is presented in Refs. [18–20]. Then, we show that our framework can classify gapless points in magnetic space groups $P21'$ and $P4$ as examples of noncentrosymmetric and TRS broken cases. The results are summarized in Table 4.5.

$P2/m1'$ with B_g Pairing

Let us begin by discussing a gapless point on the rotation-symmetric line, which is denoted by D^1 in Fig. 4.5a. On this particular line, as tabulated in Table 4.3, there are two irreducible representations. Reference [20] has argued that a gapless point on the rotation axis is part of a line node, although solid derivation has not been provided. Here, we show that gapless points pinned to the rotation axes are part of line nodes by using $d_1^{1,0}$ and our point-node classifications.

Suppose that we have $\mathfrak{n}^{(1)} = \mathfrak{b}^{(1)}_{D^1}$ whose band labels are $\mathfrak{P}^1_{D^1} = \mathfrak{P}^2_{D^1} = \mathfrak{P}^1_{TD^1} = \mathfrak{P}^2_{TD^1} = 1$. From Fig. 4.5a, we identify adjacent 2-cells to the 1-cell D^1. The 2-cells are invariant under IT and IC with $(IT)^2 = -1$ and $(IC)^2 = +1$, and EAZ classes on the 2-cells are class DIII. As a result, compatibility conditions among them do not exist; that is, $d_1^{1,0}(\mathfrak{n}^{(1)}) = 0$.

Since we now consider Case (B), we next ask if point nodes are stable on the 1-cell. As shown in Sect. 4.2.3, the classification of Dirac Hamiltonians is trivial, i.e., there are no stable point nodes on the 1-cell. This implies that the gapless point on the 1-cell is part of a line node extended from the 1-cell to 3-cells, which is exactly what Ref. [20] proposed. This line node is protected by one-dimensional winding number W defined by chiral symmetry on the 3-cells.

Next, we consider the mirror plane on $k_y = 0$. Suppose that there exists a gapless point on the 1-cell b in Fig. 4.1 and that we have $\mathfrak{n}^{(1)} = \mathfrak{b}^{(1)}_b$ whose band labels are $\mathfrak{P}^{\pm}_b = 1$ for irreducible representations $U^{\pm}_b(M_y) = \pm i$.[2] The 2-cells α and α_7 are adjacent to the 1-cell b, and their symmetry groups are the same as that of b. As a result, $d_1^{1,0}(\mathfrak{b}^{(1)}_b) = \mathfrak{b}^{(2)}_\alpha + \mathfrak{b}^{(2)}_{\alpha_7}$. Here, $\mathfrak{b}^{(2)}_\alpha$ ($\mathfrak{b}^{(2)}_{\alpha_7}$) is a generator of $E_1^{2,0}$ in which $\mathfrak{P}^{\pm}_\alpha$ ($\mathfrak{P}^{\pm}_{\alpha_7}$) and symmetry-related band labels are equal to 1. As discussed in Sect. 4.3.2, $d_1^{1,0}(\mathfrak{n}^{(1)}) \neq 0$ implies that the gapless point on the 1-cell b is part of gapless lines on the adjacent 2-cells. As discussed above, EAZ classes on 3-cells are class DIII for which topological invariants are not defined, and thus, there are no compatibility conditions between 2-cells and 3-cells, i.e., $d_1^{2,0} = 0$. Thus, the classification of gapless points on the 1-cell b is $\mathbb{Z}_2$, and the gapless point is part of a line node in the mirror plane. This result is consistent with the previous works [18, 19].

$P21'$ with B Pairing

Next, we discuss a gapless point on a rotation-symmetric line without inversion symmetry. Here, we use the cell decomposition shown in Fig. 4.5a. In this case, a gapless point pinned to the rotation axis is also part of a line node. Irreducible representations $U^{\beta=1,2}_{D^1}$ and their EAZ classes are tabulated in Table 4.3.

Again we consider $\mathfrak{n}^{(1)} = \mathfrak{b}^{(1)}_{D^1}$ whose band labels are $\mathfrak{N}^1_{D^1} = -\mathfrak{N}^2_{D^1} = +1$ and symmetry-related band labels are equal to 1 or -1. Unlike the case of $P2/m1'$, the 2-cells are invariant only under chiral symmetry Γ, and thus their EAZ classes are class AIII. Similar to the case of Sect. 4.3.3, this implies that compatibility conditions between the 1-cell and the adjacent 2-cells are absent, i.e., $d_1^{1,0}(\mathfrak{n}^{(1)}) = 0$. As a result, the gapless point on the 1-cell is not extended to its adjacent 2-cells. Since $(\mathfrak{N}^1_{D^1}, \mathfrak{N}^2_{D^1}) = (1, -1)$ does not coincide with any element of the point-node classification of Eq. (4.67), we conclude that a single gapless point on the 1-cell is part of a line node. This line node is also protected by the one-dimensional winding number W defined on 3-cells. Then, the appearance of the line node can be understood as follows. Suppose that we have a line node in $P2/m1'$ with B_g, as discussed above. Then, we break the inversion symmetry of $P2/m1'$ with B_g pairing, but the line node still survives since it is protected by W.

---

[2] Since the symmetry group of the mirror plane on $k_y = 0$ is a subgroup of wallpaper group $p4mm$, we can use the same cell decomposition.

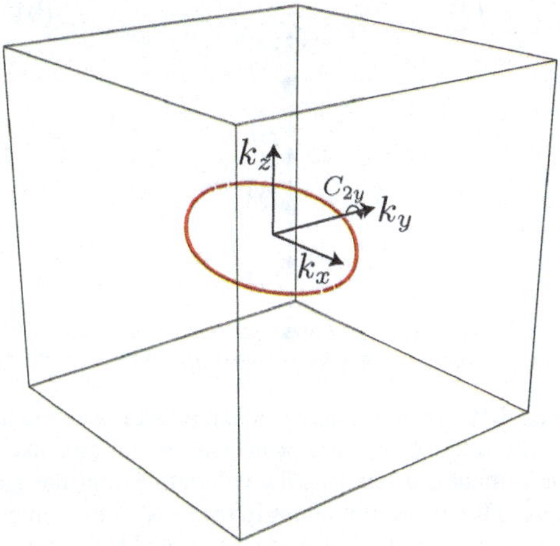

Fig. 4.9 The nodal line of the tight-binding model in Eq. (4.80) for $\mu = +1$.Reprinted from Ref. [1] by The Author(s) licensed under CC BY 4.0

We verify the existence of such line nodes through the following toy model

$$H_k = (3 - \cos k_x - \cos k_y - \cos k_z - \mu)\tau_z + (\sin k_x + 2\sin k_z)\tau_y, \tag{4.80}$$

$$\rho(C_2^y) = -i\tau_z\sigma_y, \tag{4.81}$$

$$\rho(\mathcal{T}) = i\sigma_y K, \tag{4.82}$$

$$\rho(\mathcal{C}) = \tau_x K. \tag{4.83}$$

As shown in Fig. 4.9, we find a line node.

It should be noted that $2\mathfrak{b}_{D^1}^{(1)}$ has the same band labels as Eq. (4.67). Then, it is natural to ask whether $\mathfrak{n}^{(1)} = 2\mathfrak{b}_{D^1}^{(1)}$ is a genuine point node or not. Here, we show that the above line node exists even in such a case. We start with two line nodes generated by $\mathfrak{n}^{(1)} = 2\mathfrak{b}_{D^1}^{(1)}$ illustrated in Fig.4.10a. The winding numbers can be canceled by rotating one of the line nodes, and then we have two pairs of point nodes in Fig. 4.10b. In fact, this corresponds to a generator of the classification of Dirac Hamiltonians discussed in Sect. 4.2.3. However, no symmetry constraint forces two gapless points to exist at the same momentum to make up a point node. As shown in Sect. 4.2.3, symmetric perturbations deform a point node to a shrinkable loop node. Each pair of point nodes again forms a shrinkable loop node illustrated in Fig. 4.10c. Therefore, gapless points on the 1-cell are classified into $\mathbb{Z}$, and they are part of line nodes.

$P4$ with ${}^1E$ Pairing

Next, we discuss magnetic space group $P4$ with ${}^1E$ pairing, whose generators are fourfold rotation symmetry C_4^z and PHS with $\rho(\mathcal{C})\rho(C_4^z) = i\rho(C_4^z)\rho(\mathcal{C})$. We consider 1-cell D^1 illustrated in Fig. 4.11a. Here, we show that a gapless point on the 1-cell is part of surface nodes. Table 4.4 show irreducible representations $U_{D^1}^{\beta}$ ($\beta = 1, 2, 3, 4$) and their EAZ classes.

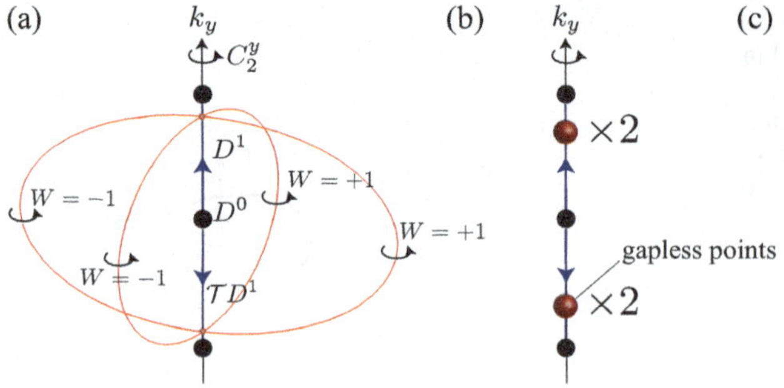

Fig. 4.10 Deformation of nodal structures in magnetic space group $P21'$. (a) Two line nodes fixed to the rotation axis are protected by one-dimensional winding numbers. (b) These line nodes are deformable to point nodes without closing the gap at 0-cells. (c) Since no symmetry constraint forces the two point nodes in (b) to be at the same momentum, each of the two point nodes in turn becomes part of a line node. Reprinted from Ref. [1] by The Author(s) licensed under CC BY 4.0

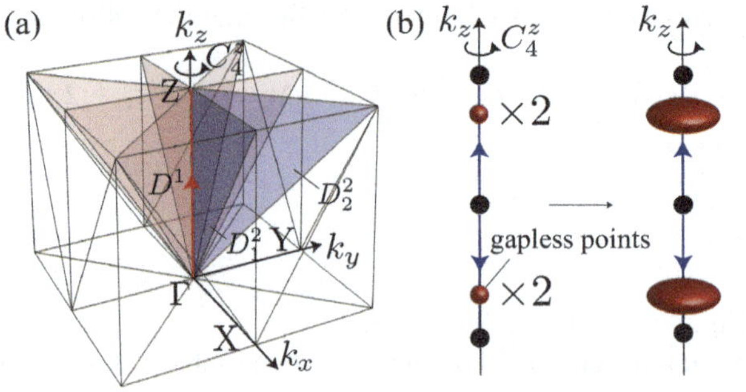

Fig. 4.11 **a** A half BZ in magnetic space group $P4$. Here, the blue planes D_1^2 and D_2^2 are adjacent 2-cells to D^1 and red ones are symmetry-related to D_1^2 and D_2^2. **b** Deformation of nodal structures in magnetic space group $P4$. Reprinted from Ref. [1] by The Author(s) licensed under CC BY 4.0

Suppose we have $\mathfrak{n}^{(1)} = \mathfrak{b}^{(1)}_{D^1,\beta=1}$ whose band labels are $\mathfrak{N}^1_{D^1} = -\mathfrak{N}^3_{CD^1} = +1$. Only two of eight adjacent 2-cells to D^1 are symmetry-inequivalent, and here we choose blue planes D_1^2 and D_2^2 in Fig. 4.11a as them. Since the EAZ classes on D^1, the adjacent 2-cells, and 3-cells are all class A, there exist compatibility conditions among them. As a result, $d_1^{1,0}(\mathfrak{n}^{(1)}) = \mathfrak{b}^{(2)}_{D_1^2} - \mathfrak{b}^{(2)}_{D_2^2}$, in which $\mathfrak{N}_{D^2_{i=1,2}} = +1$ and symmetry-related band labels are equal to 1 or -1. Furthermore, $d_1^{2,0}(\mathfrak{b}_{D_1^2}) \neq 0$ and $d_1^{2,0}(\mathfrak{b}_{D_2^2}) \neq 0$. This implies that a gapless point on the 1-cell is part of surface nodes. Although here we consider only the case of $\beta = 1$, discussions and results for other values of β are the same.

As shown in an example of Sect. 4.2.3, when $\mathfrak{n}^{(1)}$ is a linear combination of $\{\mathfrak{b}^{(1)}_{D^1,\beta}\}_{\beta=1}^4$, for example $\mathfrak{n}^{(1)} = \mathfrak{b}^{(1)}_{D^1,\beta=1} - \mathfrak{b}^{(1)}_{D^1,\beta=2}$, point nodes can exist on the 1-cell. However, the same logic in Sect. 4.3.3 is applicable again; that is, symmetric perturbation deforms a point node to a shrinkable surface node as discussed in Sect. 4.2.3.

Thus, the point nodes can be inflated as shown in the right panel of Fig. 4.11b, gapless points on the 1-cell are part of surface nodes (Bogoliubov Fermi surfaces). Ref. [5] has discussed Bogoliubov Fermi surfaces in multibands superconductors without inversion symmetry, although Ref. [5] has not discussed the symmetry-protection of them.

4.4 Further Extension for Nodes at Generic Points

Thus far, we have focused only on nodes pinned to 1-cells. However, not all nodes are the case. In this section, we discuss a possible refinement of our symmetry-based framework to cover nodes not pinned to 1-cells through the mirror plane in magnetic space group $P2/m1'$ with B_u pairing.

Here, we decompose the mirror plane in the same way in Fig. 4.1 and discuss 1-cell b of Fig. 4.1. Applying our point-node classification to the 1-cell, we find that the point-node on the 1-cell b is classified into $\mathbb{Z}$. The Dirac Hamiltonian is

$$H_{(k_1,k_2)} = k_1\tau_y + k_2\tau_x\sigma_z + \delta k_3\tau_z, \tag{4.84}$$

$$\sigma(IC) = i\tau_y K \tag{4.85}$$

$$\sigma(IT) = i\tau_z\sigma_y K \tag{4.86}$$

$$\sigma(M_y) = i\tau_z\sigma_x, \tag{4.87}$$

where k_1 and k_2 are displacement perpendicular to 1-cell b, and δk_3 is a displacement from the gapless point along 1-cell b. The point node is protected by the mirror winding number and locally stable [6]. On the other hand, since EAZ classes of mirror sectors are class AIII, no topological invariants force the point node to be on the 1-cell. This is not a contradiction. The absence of zero-dimensional topological invariants simply means that no gapless points are fixed to the 1-cell. In fact, a symmetric perturbation term shifts the point node slightly in the k_2-direction. Therefore, point nodes can locally exist everywhere and freely move in the mirror plane.

The natural question is whether these point nodes are globally stable or not. Here, we show that only two point nodes are globally stable in the mirror plane. To illustrate this, let us consider that there exist four point nodes in the plane, as shown in Fig. 4.12. Since for B_u pairing, C_2^y anticommutes with PHS, C_2^y changes the signs of mirror winding numbers (see Appendix 3). As discussed above, the point nodes can freely move within the mirror plane. Consequently, two point nodes with opposing mirror winding numbers can meet and be pair-annihilated. This suggests that a single pair of point nodes is globally stable. This annihilation process can be interpreted as "second differential" $d_2^{p,0}$ of Atiyah-Hirzebruch spectral sequence [2]. Then, it might be possible to refine our framework by incorporating the second differentials into the classification scheme.

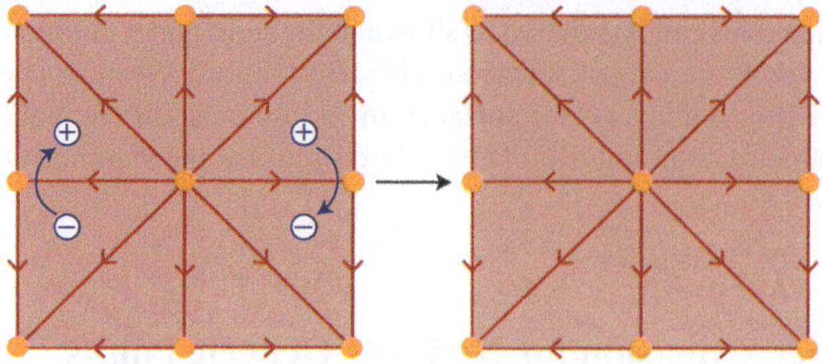

Fig. 4.12 Annihilation process of gapless points. Here white solid circles denote gapless points and ±-signs represent the sign of the winding numbers. Reprinted from Ref. [1] by The Author(s) licensed under CC BY 4.0

Finally, we comment on the relation between this point node and symmetry indicators. In this symmetry setting, symmetry indicators detect the globally stable point nodes. The symmetry indicator group is $(\mathbb{Z}_2)^2 \times \mathbb{Z}_4$, whose $\mathbb{Z}_2$-parts originate from lower dimensional gapped topological phases. The $\mathbb{Z}_4$ index is defined by

$$z_4 = \frac{1}{4} \sum_{K \in \mathrm{TRIMs}} \left(\mathfrak{N}_K^+ - \mathfrak{N}_K^-\right) \mod 4, \tag{4.88}$$

where $\mathfrak{N}_K^\pm$ is a band label for each parity-sector at TRIM K. As discussed in Ref. [13], if the system is fully gapped, $z_4 = 1$ and 3 indicate that the mirror Chern number modulo two equals one. However, the nontrivial mirror Chern numbers are forbidden in this symmetry setting [7]. This contradiction is resolved by the existence of point nodes in the mirror plane. In fact, $z_4 = 1$ and 3 indicate such point nodes.

4.5 Conclusions of This Chapter

In this chapter, we have proposed a systematic framework for classifications of gapless points pinned to arbitrary lines in momentum space. Our method is mainly based on three ingredients: (i) decomposition of momentum space into points (0-cells), line segments (1-cells), planes (2-cells), and polyhedrons (3-cells), (ii) compatibility conditions, and (iii) point-node classifications on 1-cells.

As discussed in the next chapter, our theory provides a practical way to diagnose the superconducting nodes of realistic materials. In Chap. 5, we discuss an integration of our algorithm into DFT calculations, which enables a high-throughput diagnosis of superconducting nodes of materials in the database. Such studies could reduce candidate pairing symmetries of unconventional superconductors compatible with experimental observations. We believe that our theory helps to deeply understand superconductors ever discovered and synthesized in the future. As a demonstration, we analyze the superconducting nodes of several recently discovered superconductors by assuming various possible pairing symmetries.

Our theory opens up various possibilities for future studies. Although the presented framework covers a wide range of nodes, it fails to classify nodes not pinned to 1-cells, as discussed in Sect. 4.4. As mentioned in Sect. 4.4, our symmetry-based approach has a chance to be more refined, and we leave a complete understanding of relations between symmetry indicators and the nodes as a future work. Such a study would also give us more information about nodes pinned to 1-cells as follows. Suppose that a system violates compatibility conditions between 0- and 1-cells. In such a case, gapless points must be on 1-cells as discussed in this chapter. Since we can always forget about symmetries that impose the violated compatibility conditions on the system, we can consider the system as that with a lowered symmetry group. Then, we can compute symmetry indicators for the lower symmetry setting, and symmetry indicators would clarify topological nature behind the nodes as discussed in Ref. [8].

Although we have applied our framework only to superconductors, it is also applicable to normal conducting phases. Indeed, recent topological material databases [9–12] contain topological semimetals whose irreducible representations violate compatibility conditions on high-symmetry lines. Thus, the application of our framework to them could provide more fruitful information on nodes. We leave this as a future study.

Appendix 1: Cell Decomposition for Representative Space Groups

In this section, we present fundamental domains that can fill the entire momentum space through symmetry operations. When we consider a magnetic space group with at least one supergroup, we can use the cell decomposition for the supergroup as the cell decomposition for the original magnetic space group. Therefore, it is enough to have the cell decomposition for magnetic space groups $Pm\bar{3}m1'$, $Cmmm1'$, $Amm21'$, $R\bar{3}m11'$, $P6/mmm1'$, $Fm\bar{3}m1'$, and $Im\bar{3}m1'$, which can be used for all magnetic space groups with the same primitive lattice vectors. It should be noted that the cell decomposition for $Amm21'$ is the same as that for $Cmmm1'$ with axes exchanged and that the cell decomposition for $R\bar{3}m11'$ is constructed by $\{k_{j,x}\boldsymbol{b}_1 + k_{j,y}\boldsymbol{b}_2 + k_{j,z}\boldsymbol{b}_3\}_j$, where $(k_{j,x}, k_{j,y}, k_{j,z})$ is a cell for $Pm\bar{3}m1'$ and $\boldsymbol{b}_i$ is a primitive reciprocal lattice vector of $R\bar{3}m11'$. Most importantly, they satisfy all conditions discussed in Sect. 4.1.1.

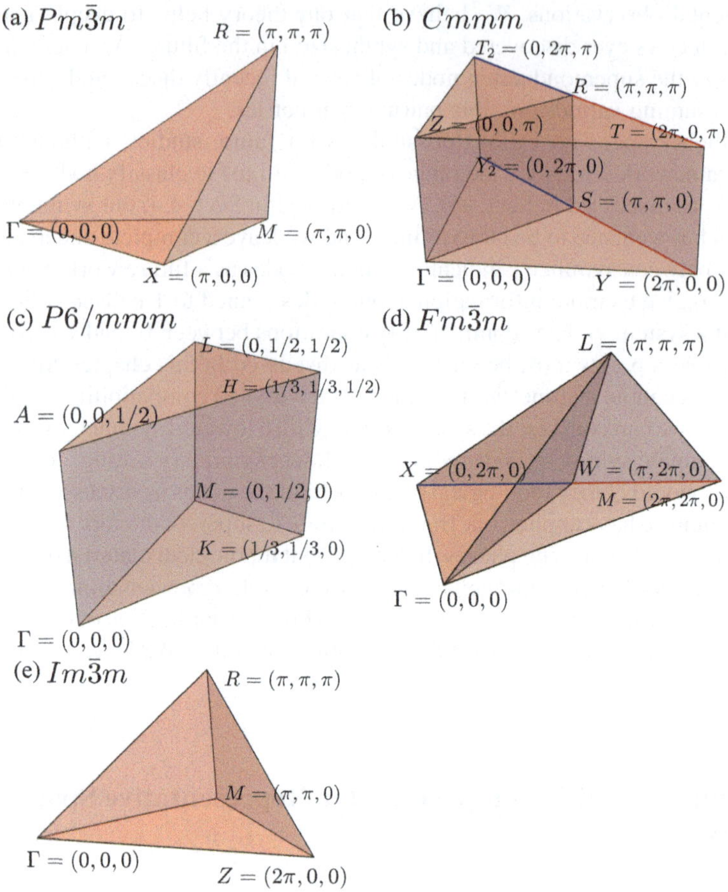

Fig. 4.13 Fundamental domains, which can span the entire momentum space by symmetry operations, for magnetic space groups $Pm\bar{3}m1'$ (**a**), $Cmmm1'$ (**b**), $P6/mmm1'$ (**c**), $Fm\bar{3}m1'$ (**d**), and $Im\bar{3}m1'$ (**e**). For $P6/mmm$, coordinates denote coefficients of primitive reciprocal lattice vectors. Orientations except for blue and red lines in $Cmmm1'$ and $Fm\bar{3}m1'$ can be arbitrarily chosen. Orientations of these colored lines are chosen in symmetric manners. Reprinted from Ref. [1] by The Author(s) licensed under CC BY 4.0

Appendix 2: Derivation Other Formulas

In Chap. 4, we presented formulas (Table 4.2) that enable us to obtain elements of $E_1^{1,0}$ from generating Dirac Hamiltonians. In this section, we provide Gamma matrices $\gamma_{0,1,2}$ and symmetry representations $\widetilde{\sigma}(g)$ of generators of Dirac Hamiltonians for each EAZ class. By substituting them into Eqs. (4.46) and (4.47), we can derive the formulas listed in Table 4.2 (Fig. 4.13).

Table 4.6 Gamma matrices in Eq. (4.28) **and onsite unitary symmetries.** Here, $\widetilde{u}^{\widetilde{\alpha}}$ is an irreducible representation of the onsite unitary symmetry group $\widetilde{\mathcal{G}}$. In addition, $\mathcal{T}\widetilde{\alpha}$, $\mathcal{C}\widetilde{\alpha}$, and $\Gamma\widetilde{\alpha}$ are labels of irreducible representations related to $\widetilde{\alpha}$ by onsite time-reversal, particle-hole, and chiral symmetry, respectively. Reprinted from Ref. [1] by The Author(s) licensed under CC BY 4.0

EAZ	γ_0	γ_1	γ_2	$\widetilde{\sigma}(g)$
A	σ_z	σ_x	σ_y	$\mathbb{1} \otimes \widetilde{u}^{\widetilde{\alpha}}(g)$
$\mathrm{A}_{\mathcal{T}}$	τ_z	τ_x	$\tau_y\sigma_z$	$\tau_0 \mathrm{diag}(\widetilde{u}^{\widetilde{\alpha}}(g), \widetilde{u}^{\mathcal{T}\widetilde{\alpha}}(g))$
$\mathrm{A}_{\mathcal{C}}$	τ_z	τ_x	τ_y	$\tau_0 \mathrm{diag}(\widetilde{u}^{\widetilde{\alpha}}(g), \widetilde{u}^{\mathcal{C}\widetilde{\alpha}}(g))$
A_{Γ}	τ_z	τ_x	$\tau_y\sigma_z$	$\tau_0 \mathrm{diag}(\widetilde{u}^{\widetilde{\alpha}}(g), \widetilde{u}^{\Gamma\widetilde{\alpha}}(g))$
$\mathrm{A}_{\mathcal{T},\mathcal{C}}$	τ_z	τ_x	$\tau_y(\mathbb{1} \otimes \sigma_z)$	$\tau_0 \mathrm{diag}(\widetilde{u}^{\widetilde{\alpha}}(g), \widetilde{u}^{\mathcal{T}\widetilde{\alpha}}(g), \widetilde{u}^{\mathcal{C}\widetilde{\alpha}}(g), \widetilde{u}^{\Gamma\alpha}(g))$
C	σ_z	σ_x	σ_y	$\mathbb{1} \otimes \widetilde{u}^{\widetilde{\alpha}}(g)$
$\mathrm{C}_{\mathcal{T}}$	τ_z	τ_y	$\tau_x\sigma_z$	$\tau_0 \mathrm{diag}(\widetilde{u}^{\widetilde{\alpha}}(g), \widetilde{u}^{\mathcal{T}\widetilde{\alpha}}(g))$
D	τ_z	$\tau_x\sigma_y$	$\tau_y\sigma_y$	$\tau_0 \mathrm{diag}(\widetilde{u}^{\widetilde{\alpha}}(g), \widetilde{u}^{\widetilde{\alpha}}(g))$
$\mathrm{D}_{\mathcal{T}}$	s_z	$s_y\tau_y$	$s_x\tau_y\sigma_z$	$s_0\tau_0 \mathrm{diag}(\widetilde{u}^{\widetilde{\alpha}}(g), \widetilde{u}^{\mathcal{T}\widetilde{\alpha}}(g))$
AI	σ_z	σ_x	$\tau_y\sigma_y$	$\tau_0 \mathrm{diag}(\widetilde{u}^{\widetilde{\alpha}}(g), \widetilde{u}^{\widetilde{\alpha}}(g))$
$\mathrm{AI}_{\mathcal{C}}$	s_z	s_x	$s_y\tau_z$	$s_0\tau_0 \mathrm{diag}(\widetilde{u}^{\widetilde{\alpha}}(g), \widetilde{u}^{\mathcal{C}\widetilde{\alpha}}(g))$
CI	τ_z	τ_x	$\tau_y\sigma_z$	$\tau_0 \mathrm{diag}(\widetilde{u}^{\widetilde{\alpha}}(g), \widetilde{u}^{\widetilde{\alpha}}(g))$

Appendix 3: Symmetry Property of the Winding Number

In this section, we show that C_2^y that holds $\rho(C_2^y)\rho(\mathcal{C}) = -\rho(\mathcal{C})\rho(C_2^y)$ changes the sign of the one-dimensional winding number. The winding number is defined by (Table 4.6)

$$W[C] \equiv \oint_C \mathrm{tr}[U(\Gamma)\left(H_k^{-1}\partial_k H_k\right)]ds, \tag{4.89}$$

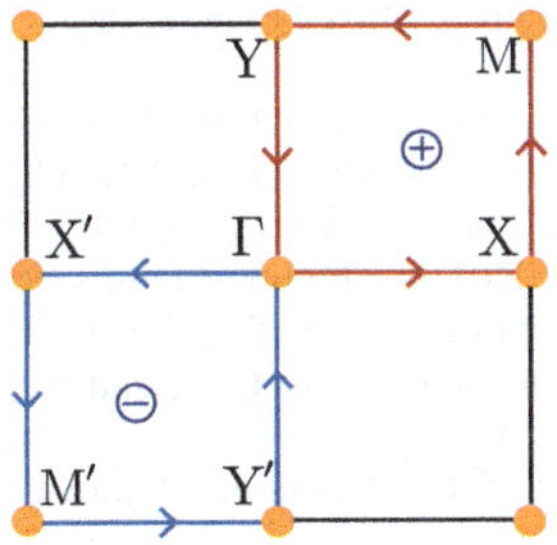

Fig. 4.14 Interval of integration used in calculating the winding number. It is important to note that the path marked in blue is symmetry-related to the one in red. Here, solid white circles represent gapless points with the winding number ±1, and the symbols ⊥ in the circles denote their signs. Reprinted from Ref. [1] by The Author(s) licensed under CC BY 4.0

where we take the path $\Gamma - \mathrm{X} - \mathrm{M} - \mathrm{Y} - \Gamma$ as C (see Fig. 4.14). We first compute the integrand

$$\begin{aligned}
&\mathrm{tr}[U(\Gamma)\left(H_{\boldsymbol{k}}^{-1}\partial_{k_i}H_{\boldsymbol{k}}\right)] \\
&= \mathrm{tr}[U(C_2^y)U(\Gamma)\left(H_{\boldsymbol{k}}^{-1}\partial_{k_i}H_{\boldsymbol{k}}\right)U^{-1}(C_2^y)] \\
&= -\mathrm{tr}[U(\Gamma)U(C_2^y)\left(H_{\boldsymbol{k}}^{-1}\partial_{k_i}H_{\boldsymbol{k}}\right)U^{-1}(C_2^y)] \\
&= -\mathrm{tr}[U(\Gamma)\left(H_{-\boldsymbol{k}}^{-1}\frac{\partial}{\partial k_i}H_{-\boldsymbol{k}}\right)].
\end{aligned} \tag{4.90}$$

Using this relation, we can rewrite an integral for a part of the path

$$\begin{aligned}
&\int_{\Gamma}^{\mathrm{X}} \mathrm{tr}[U(\Gamma)\left(H_{(k_x,0)}^{-1}\partial_{k_x}H_{(k_x,0)}\right)]dk_x \\
&= -\int_{\Gamma}^{\mathrm{X}'} \mathrm{tr}[U(\Gamma)\left(H_{(k_x,0)}^{-1}\partial_{k_x}H_{(k_x,0)}\right)]dk_x
\end{aligned} \tag{4.91}$$

For other parts, we have the same transformations. As a result, we find the relation $W[C] = -W[C_2^y C]$. Indeed, the relation can be generalized to other point group symmetries as $W[C] = \chi_g \det p_g W[gC]$ ($\chi_g = \pm 1$).

References

1. Ono S, Ken S (2022) Symmetry-based approach to superconducting nodes: unification of compatibility conditions and gapless point classifications. Phys Rev X 12:011021
2. Ken S, Masatoshi S, Kiyonori G (2022) Atiyah-Hirzebruch spectral sequence in band topology: General formalism and topological invariants for 230 space groups. Phys. Rev. B 106:165103
3. Po HC, Watanabe H, Vishwanath A (2018) Fragile topology and Wannier obstructions. Phys Rev Lett 121:126402
4. Elcoro L, Bradlyn B, Wang Z, Vergniory MG, Cano J, Felser C, Andrei Bernevig B, Orobengoa D, de la Flor G, Aroyo MI (2017) . Double crystallographic groups and their representations on the Bilbao Crystallographic Server. J Appl Crystallogr 50(5):1457–1477
5. Link Julia M, Herbut Igor F (2020) Bogoliubov-Fermi surfaces in noncentrosymmetric multicomponent superconductors. Phys Rev Lett 125:237004
6. Yang Shengyuan A, Hui P, Zhang Fan (2014) Dirac and Weyl superconductors in three dimensions. Phys Rev Lett 113:046401
7. Zhang F, Kane CL, Mele EJ (2013) Topological mirror superconductivity. Phys Rev Lett 111:056403
8. Zhang T, Lu L, Murakami S, Fang Z, Weng H, Fang C (2020) Diagnosis scheme for topological degeneracies crossing high-symmetry lines. Phys Rev Res 2:022066(R)
9. Zhang T, Jiang Y, Song Z, Huang H, He Y, Fang Z, Weng H, Fang C (2019) Catalogue of topological electronic materials. Nature 566(7745):475–479
10. Vergniory MG, Elcoro L, Felser C, Regnault N, Andrei Bernevig B, Wang Z (2019) A complete catalogue of high-quality topological materials. Nature 566(7745):480–485

11. Tang F, Po HC, Vishwanath A, Wan X (2019) Comprehensive search for topological materials using symmetry indicators. Nature 566(7745):486–489
12. Xu Y, Elcoro L, Song Z-D, Wieder BJ, Vergniory MG, Regnault N, Chen Y, Felser C, Andrei Bernevig B (2020) High-throughput calculations of magnetic topological materials. Nature 586(7831):702–707
13. Ono S, Po HC, Watanabe H (2020) Refined symmetry indicators for topological superconductors in all space groups. Sci Adv 6(18):eaaz8367
14. Ryu S, Schnyder AP, Furusaki A, Ludwig AWW (2010) Topological insulators and superconductors: tenfold way and dimensional hierarchy. New J Phys 12(6):065010
15. Cornfeld E, Chapman A (2019) Classification of crystalline topological insulators and superconductors with point group symmetries. Phys Rev B 99:075105
16. Ken S (2022) The classification of surface states of topological insulators and superconductors with magnetic point group symmetry. Progr Theoret Experim Phys 2022(4):4A104
17. Teo JCY, Kane CL (2010) Topological defects and gapless modes in insulators and superconductors. Phys Rev B 82:115120
18. Sumita S, Yanase Y (2018) Unconventional superconducting gap structure protected by space group symmetry. Phys Rev B 97:134512
19. Kobayashi S, Sumita S, Yanase Y, Sato M (2018) Symmetry-protected line nodes and Majorana flat bands in nodal crystalline superconductors. Phys Rev B 97:180504(R)
20. Sumita S, Nomoto T, Shiozaki K, Yanase Y (2019) Classification of topological crystalline superconducting nodes on high-symmetry lines: point nodes, line nodes, and bogoliubov fermi surfaces. Phys Rev B 99:134513

Chapter 5
High-Throughput Investigation

So far, we have developed theoretical frameworks to diagnose topological and nodal superconductivity based on band topology. As mentioned in Chap. 1, investigating nodal and topological superconductivity of realistic materials is one of the important steps to elucidate the mechanisms of superconductivity in unconventional superconductors. Unfortunately, experimentally confirmed nodal and topological superconductors are significantly scarce compared to topological insulators and semimetals. Given the success of Refs. [1–4], one might think that the establishment of such a database for superconductors would significantly boost the search for new topological superconductor candidates. However, because pairing symmetries of actual materials are usually unknown; consequently, this attempt is quite challenging.

In this chapter, we avoid dealing with this difficulty and take an alternative approach: We apply the theories developed in Chaps. 3 and 4 to materials by assuming various possible pairing symmetries and build a database summarizing the topological and nodal superconducting nature of the materials for the assumed pairings.[1] We suppose that all the computed materials become superconductors at low temperatures. It should be emphasized that we do not aim to predict topological and nodal superconductivity for an actual superconducting pairing. Even if the actual pairing symmetry has not been identified yet, applying our theories to each possible pairing symmetry is still helpful in clarifying the mechanisms of unconventional superconductivity. These predictions could be compared with experimental observations and exclude pairing symmetries inconsistent with experiments from the possibilities. In addition to the database, we provide a subroutine *Topological Supercon* that enables an automatic diagnosis of topological or nodal superconducting nature of any material synthesized in the future.[2]

[1] All computational results are listed in our database *Database of Topological and Nodal Superconductors* (https://tnsc.nju.edu.cn).

[2] This subroutine can diagnose possible topological or nodal superconductivity under various conditions that are not covered in the database, for example, in the absence of spin-orbit coupling.

S. Ono, *Topology of Superconductors*, Springer Theses,
https://doi.org/10.1007/978-981-97-7399-2_5

5.1 Diagnostic Scheme

In this section, we provide our diagnostic scheme illustrated in Fig. 5.1. Generally speaking, the full diagnosis of topology requires (i) Bloch wave functions over the entire BZ and (ii) the complete list of topological indices to be computed. However, it is usually difficult to obtain complete information on Bloch wave functions for realistic materials, especially for superconductors. In addition, the complete sets of topological invariants are also unknown for most magnetic space groups. In view of this, symmetry-indicator theory, in which we focus only on zero-dimensional topological invariants of systems, enables us to efficiently diagnose topological and nodal nature of the superconducting system.

As discussed in Chap. 3, zero-dimensional topological invariants form a list $\vec{n}_{SC}$. The only problem is to obtain $\vec{n}_{SC}$ for target materials. To get zero-dimensional topological invariants for realistic materials, we make an assumption, so-called *weak pairing assumption*, that the energy scale of the pair potential in superconducting phases is typically much smaller than that of normal conducting phases [6]. Under this assumption, once we assume a pairing symmetry, zero-dimensional topological invariants in the superconducting phase can be deduced from the numbers of

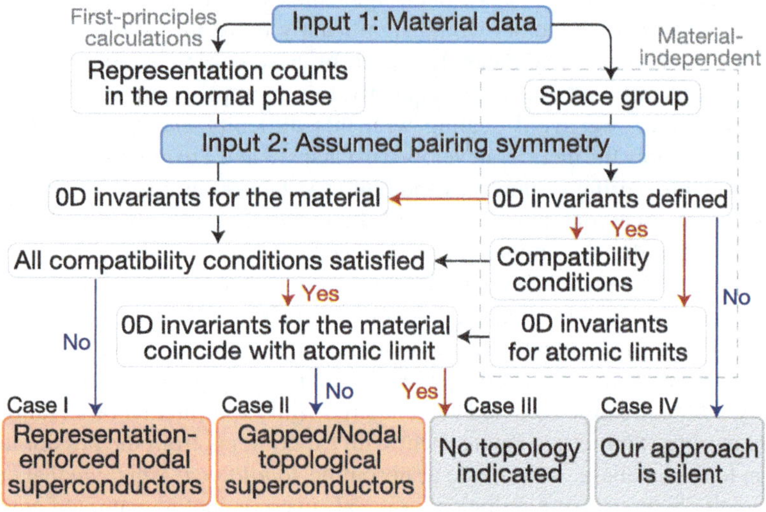

Fig. 5.1 **Flow chart of our diagnostic scheme.** For a given material, the initial step is to identify its space group and the irreducible representations of its normal phase. Once the space group is specified, assuming a pairing symmetry, we know whether zero-dimensional topological invariants can be defined or not. If any invariant is not definable, our approach makes no predictions. If they are, we proceed to tabulate compatibility conditions and zero-dimensional topological invariants for atomic limits, and we can diagnose the topological and nodal superconducting nature of the material based on them. Reprinted from Ref. [5] by The Author(s) licensed under CC BY 4.0 with some modification

occupied bands characterized by irreducible representations in the normal phase.[3] More concretely, we obtain zero-dimensional topological invariants defined in Eqs. (3.27) and (3.28) by the following formulas:

$$P_{\boldsymbol{k}}^{\alpha} \to n_{\boldsymbol{k}}^{\alpha}|_{\text{occ}} \mod 2, \tag{5.1}$$

$$N_{\boldsymbol{k}}^{\alpha} \to n_{\boldsymbol{k}}^{\alpha}|_{\text{occ}} - n_{-\boldsymbol{k}}^{\bar{\alpha}}|_{\text{occ}}, \tag{5.2}$$

where $n_{\boldsymbol{k}}^{\alpha}|_{\text{occ}}$ represents the number of occupied bands with an irreducible representation $U_{\boldsymbol{k}}^{\alpha}$. Also, $n_{-\boldsymbol{k}}^{\bar{\alpha}}|_{\text{occ}}$ is a count of occupied bands with the particle-hole conjugate irreducible representation for $U_{\boldsymbol{k}}^{\alpha}$, i.e., $U_{-\boldsymbol{k}}^{\bar{\alpha}}(h) = \frac{z_{h,C}}{z_{C,C^{-1}hC}}[U_{\boldsymbol{k}}^{\alpha}(C^{-1}hC)]^{*}$.

Since zero-dimensional topological invariants must not change during any continuous deformation, zero-dimensional topological invariants in $\vec{n}_{\text{SC}}$ for a gapped Bogoliubov quasiparticle spectrum are not fully independent and are subject to compatibility conditions. If some compatibility conditions are violated, the superconductor is a nodal superconductor, to which we refer as *representation-enforced nodal superconductor*. Violations of compatibility conditions ensure that gapless points must be on lines connecting two high-symmetry momenta. Then, by checking the classification results discussed in Chap. 4, we can determine the positions of gapless points and their dimensionality such as point, line, and surface. When all compatibility conditions are satisfied, we can apply our symmetry-indicator theory to $\vec{n}_{\text{SC}}$. In other words, we compare $\vec{n}_{\text{SC}}$ with sets of zero-dimensional topological invariants for topologically trivial superconductors that are actually atomic limit ones introduced in Chap. 3. A superconductor whose $\vec{n}_{\text{SC}}$ is different from any atomic limit must be topologically nontrivial, i.e., the superconductor is a topological superconductor.[4]

Applying the above scheme to the obtained zero-dimensional topological invariants, we analyze topological and nodal superconducting nature. There are four cases of results: representation-enforced nodal superconductor (Case I); symmetry-diagnosable topological or nodal superconductor (Case II); topologically trivial or not symmetry-diagnosable topological superconductor (Case III); the case where any zero-dimensional topological invariant cannot be defined at any high-symmetry momenta (Case IV).When TRS exists, the result for the trivial pairing symmetry of each space group is always Case IV.

[3] Although the pairing symmetry is an input of our diagnostic scheme, our framework does not require information on which degrees of freedom, such as spins and orbitals, pair up. In other words, our analysis focuses only on symmetry properties of Cooper pairs and does not distinguish between pairing mechanisms. This makes the analysis more accessible and systematic.

[4] It should be noted that band structures satisfying all compatibility conditions are not always gapped. As discussed in Refs. [7, 8], symmetry indicators sometimes detect gapless phases with topological origins.

5.2 Material Examples

In this section, we apply our framework to some example materials. The results of our symmetry-based analysis are tabulated in Table 5.1. Detailed computations for several examples are shown in Appendix

5.2.1 *CaPtAs*

First, we apply the above diagnostic scheme to CaPtAs, whose space group is $I4_1md$ (No. 109). Point-node superconductivity and time-reversal symmetry breaking have been reported in Ref. [9]. A possible pairing symmetry that breaks TRS is the two-dimensional irreducible representation E. The order parameter that belongs to the two-dimensional irreducible representation also breaks some crystalline symmetries, i.e., the space group is reduced to space group $I4_1$ (No. 80). There are only two possibilities compatible with time-reversal symmetry breaking: ${}^1E$ ($\chi_{C_4} = -i$) and ${}^2E$ ($\chi_{C_4} = i$) representations of the point group C_4. Since the case for ${}^2E$ is the same as that for ${}^1E$, we focus only on the latter case.

As the first step, we perform DFT calculations and obtain irreducible representations of space group $I4_1md$. Since we can always decompose irreducible representations in $I4_1md$ into those in $I4_1$, we get zero-dimensional topological invariants for $I4_1$ by assuming that the pairing symmetry is ${}^1E$ representation and that the weak-pairing assumption holds. After examining whether all compatibility conditions are satisfied or not, we find two violated compatibility conditions on the rotation symmetric line between $\Gamma = (0, 0, 0)$ and $\mathrm{M} = (0, 0, 2\pi)$. In fact, the symmetry setting on this line is completely the same as that in Sect. 4.3.3, and thus the nodal structures are also the same. Since the two gapless points correspond to a set of zero-dimensional topological invariants listed in Eq. (4.75), we expect that this material has small Bogoliubov Fermi surfaces as discussed in Sect. 4.3.3 (see Fig. 4.11b).

Our result and the experimental observation of point nodes might not contradict each other. Since the superconducting order parameters are considered to be much smaller in most superconductors, it is reasonable to think that the Bogoliubov Fermi surfaces are also small. To distinguish between this case and *exact* point nodes, further experiments are required.

5.2.2 *BaPtP*

Next, we discuss superconducting nature of BaPtP, whose superconductivity was recently reported [10]. The space group of BaPtP is $P2_13$ (space group No. 198), which lacks inversion symmetry [11]. This material is predicted to host unconventional excitations with four and sixfold degeneracies [1, 12]. Interestingly, the con-

Table 5.1 Results of symmetry-based diagnosis for CaPtAs, BaPtP, BaPtSb, and CsV_3Sb_5

Material	PG	Irrep $\mathcal{D}(g)$	Case	Nodes and Topology
CaPtAs (ICSD: 060828)	C_4	${}^2E$	I [S]	Γ-M
		${}^1E$	I [S]	Γ-M
	C_{4v}	B_2	I [P]	Γ-M
		B_1	I [P]	Γ-M
		A_2	II	(0, 1, 1)
		A_1	IV	–
BaPtP (ICSD: 059191)	T	${}^2E$	I [S]	Γ-R
		${}^1E$	I [S]	Γ-R
		A	IV	–
	D_2	B_1	I [L]	Γ-Z
		B_2	I [L]	Γ-Y
		B_3	I [L]	Γ-X
BaPtSb (ICSD: 059186)	D_{3h}	A_2''	I [L]	K-H
		A_2'	I [L]	Γ-K, M-K, K-H
		A_1''	II	(0, 1, 0, 0, 5, 5)
		A_1'	IV	–
CsV_3Sb_5 Star of David	D_{6h}	B_{1u}	I [P]	Γ-A
		B_{1g}	I [L]	Γ-A
		B_{2u}	I [P]	Γ-A
		B_{2g}	I [L]	Γ-A
		A_{2u}	I [P]	Γ-A
		A_{2g}	I [L]	Γ-A
		A_{1u}	II	(0, 0, 0, 1, 8, 8)
		A_{1g}	IV	–
CsV_3Sb_5 Tri-Hexagonal	D_{6h}	B_{1u}	I [P]	A-L
		B_{1g}	I [L]	Γ-K, M-K, A-L, L-H, Γ-A
		B_{2u}	I [P]	Γ-K, M-K, L-H
		B_{2g}	I [L]	Γ-K, M-K, A-L, L-H, Γ-A
		A_{2u}	I [P]	Γ-A
		A_{2g}	I [L]	Γ-K, M-K, A-L, L-H, Γ-A
		A_{1u}	II	(0, 0, 1, 0, 9, 11)
		A_{1g}	IV	–

The table is structured as follows: the first column lists the names of materials. The second and third columns specify the point group in the superconducting phase and the assumed pairing symmetries. The fourth column tabulates the diagnostic results, Cases I–IV. For Case I, the symbols [S], [L], and [P] indicate the dimensionality of the nodes, namely surface, line, and point, respectively. The last column shows the lines where compatibility conditions are violated for Case I, and an entry of the symmetry indicator group for Case II. Specifically, for Case II of CaPtAs, BaPtSb, and CsV_3Sb_5, the entries correspond to $(\mathbb{Z}_2)^3$, $(\mathbb{Z}_2)^4 \times (\mathbb{Z}_6)^2$, and $(\mathbb{Z}_2)^4 \times \mathbb{Z}_{12} \times \mathbb{Z}_{24}$, respectively.

stituent elements belong to the same families as those in CaPtAs, although their crystal structures are different. The electronic band structure computed by first-principles calculations is shown in Fig. 5.2a.

The point group of this space group is T, in which there are three one-dimensional single-valued representations: A, ${}^1E$, and ${}^2E$. While A representation preserves TRS, the latter two break TRS. As mentioned above, the result for the A pairing is Case IV. On the other hand, for ${}^1E$ or ${}^2E$ pairing, compatibility conditions along the Γ–R line are violated. As a result, a superconductor with ${}^1E$ or ${}^2E$ pairing is a representation-enforced nodal superconductor. The resulting nodes along this line turn out to be part of a surface node.

This point group has a three-dimensional irreducible representation T. As discussed in Sect. 3.1.2, multi-dimensional irreducible representations break TRS and/or some spatial symmetries. Indeed, this representation can be decomposed into $B_1 \oplus B_2 \oplus B_3$ of point group D_2 or $A \oplus^1 E \oplus^2 E$ of point group C_3. When either one of B_1, B_2, or B_3 of D_2 is assumed, space group $P2_13$ is reduced to $P2_12_12_1$ (No. 19). Similar to the above case, we can obtain irreducible representations of occupied bands in the reduced space group $P2_12_12_1$ from those in the original space group $P2_13$. Then, we can perform the same analyses. As a result, we find the superconducting phase exhibits nodal lines.

5.2.3 CsV_3Sb_5

Finally, we apply our diagnostic scheme to a new kagome metal CsV_3Sb_5, which has been recently synthesized and attracted much attention because this material exhibits various orders such as superconductivity and charge-density wave [14, 15]. Here, we discuss two recently proposed crystal structures, "Star of David" and "Tri-Hexagonal," which are distorted due to the kagome breathing mode [13, 16] (see illustrations in Fig. 5.2g, h). Both crystal structures have space group $P6/mmm$ (No. 191), and we assume that no further symmetry breaking occurs.

The results for these structures are summarized in Table 5.1 for one-dimensional $\mathcal{D}(g)$. The predicted results are mostly representation-enforced nodal superconductors unless the pairing symmetry is A_{1u} or A_{1g}. Point and line nodes are enforced by violations of compatibility conditions in odd-parity and even-parity pairings, respectively.

By contrast, the superconductor with the A_{1u} pairing might exhibit gapped topological superconductivity. As for the Star of David structure, the nontrivial entry $(0, 0, 0, 1, 8, 8)$ of $X_{\text{BdG}} = (\mathbb{Z}_2)^4 \times \mathbb{Z}_{12} \times \mathbb{Z}_{24}$ indicates strong topological or topological crystalline superconductivity. Namely, the three-dimensional winding number must be even (but could also be zero), and the mirror Chern number is zero modulo six on the $k_z = 0$ plane and two modulo six on the $k_z = \pi$ plane. On the other hand, for the Tri-Hexagonal structure, the nontrivial entry $(0, 0, 1, 0, 9, 11)$ of $X_{\text{BdG}} = (\mathbb{Z}_2)^4 \times \mathbb{Z}_{12} \times \mathbb{Z}_{24}$ indicates that the gapped phase is a strong topological

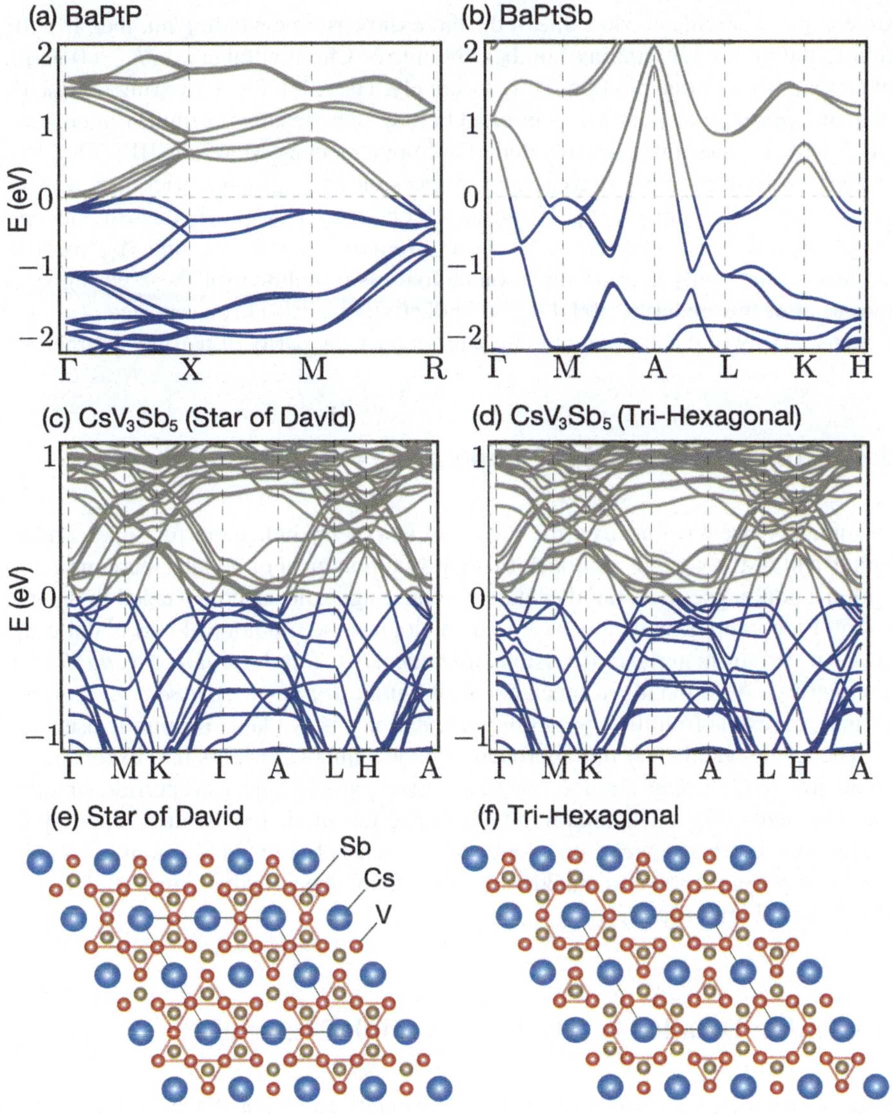

Fig. 5.2 Structures and band structures of material examples. Band structures of BaPtP (**a**); BaPtSb (**b**); CsV_3Sb_5 in 'Star of David' (**c**) and 'Tri-Hexagonal' (**d**) structures. 'Star of David' (**e**) and 'Tri-Hexagonal' (**f**) crystal structures of CsV_3Sb_5 introduced in Ref. [13]. Reprinted from Ref. [5] by The Author(s) licensed under CC BY 4.0

superconductor with an odd value of the three-dimensional winding number. In addition to the nontrivial winding number, the mirror Chern numbers on $k_z = 0$ and π planes are two and one modulo six, respectively. The result for A_{1g} paring is Case IV, and our approach is silent, as mentioned before.[5] The results for multi-dimensional irreducible representations are included in Supplementary Materials III of Ref. [5].

Our results impose strong constraints on possible pairing symmetries in the gapped phase reported recently [18]. If the superconductor is gapped, one-dimensional single-valued representations of the point group except for A_{1u} and A_{1g} must be excluded, since fully gapped superconductivity is prohibited for these paring symmetries. For this material, nodal superconductivity has also been reported [19]. The dimensionality of the nodes would help us identify the parity of pairing symmetries.

5.3 Conclusions of This Chapter

In summary, we have utilized the theory of symmetry indicators for superconductors in a high-throughput approach to identify topological or nodal superconductors within a nonmagnetic materials database. Our diagnostic scheme has led to the creation of a comprehensive database that catalogues the topological and nodal properties of various materials (accessible at *Database of Topological and Nodal Superconductors*). Additionally, a specialized subroutine *Topological Supercon*, designed to analyze results from first-principles calculations, is available for public use.

The results yielded by our approach provide valuable insights into experimental signatures, which may include boundary states, specific heat properties, or other measurements. The substantial number of materials predicted by our method opens avenues for future research into exotic states, such as Majorana fermions, a recently proposed phenomenon associated with topological superconductivity [20, 21], and Bogoliubov Fermi surfaces.

Appendix: Detailed Analysis of Example Materials

In this appendix, we show step-by-step computations of our diagnosis for several example materials. In the following, we follow the labels of irreducible representations in Ref. [22]. Also, all results of first-principles calculations are obtained from *Database of Topological and Nodal Superconductors* (https://tnsc.nju.edu.cn).

[5] Recently, it turns out that there are no gapped topological phases in this symmetry setting [17].

Table 5.2 EAZ classes at high-symmetry points of magnetic space group $P2_13$

Irrep	EAZ class	topological invariant
Γ_5	class A	$N^5_\Gamma = n^5_\Gamma - n^7_\Gamma$
Γ_6	class C	None
Γ_7	class A	$N^7_\Gamma = n^7_\Gamma - n^5_\Gamma$
X_2	class A	$N^2_X = n^2_X - n^5_X$
X_3	class A	$N^3_X = n^3_X - n^4_X$
X_4	class A	$N^4_X = n^4_X - n^3_X$
X_5	class A	$N^5_X = n^5_X - n^2_X$
M_5	class D	$P^5_M = n^5_M \mod 2$
R_4	class A	$N^4_R = n^4_R - n^6_R$
R_5	class D	$P^5_R = n^5_R \mod 2$
R_6	class A	$N^6_R = n^6_R - n^4_R$
R_7	class D	$P^7_R = n^7_R \mod 2$

BaPtP with ^1E Pairing

Here, we begin by considering BaPtP with ^1E pairing. First, according to the first-principles calculation, we find counts of irreducible representations at each high-symmetry momentum

$$(n^5_\Gamma, n^6_\Gamma, n^7_\Gamma, n^2_X, n^3_X, n^4_X, n^5_X, n^5_M, n^4_R, n^5_R, n^6_R, n^7_R) = (31, 30, 30, 45, 45, 45, 45, 90, 14, 16, 16, 46). \tag{5.3}$$

Next, based on EAZ classes and zero-dimensional topological invariants shown in Table 5.2, we obtain the following $\vec{n}_{SC}$:

$$\begin{aligned}\vec{n}_{SC} &= (N^5_\Gamma, N^7_\Gamma, N^2_X, N^3_X, N^4_X, N^5_X, P^5_M, N^4_R, P^5_R, N^6_R, P^7_R),\\ &= (1, -1, 0, 0, 0, 0, 0, -2, 0, 2, 0).\end{aligned} \tag{5.4}$$

Last, we check if the above zero-dimensional topological invariants satisfy all compatibility conditions. We find that $C\vec{n}_{SC} \neq \vec{0}$, which implies that the superconductor is a representation-enforced nodal superconductor. Nonzero elements in $C\vec{n}_{SC}$ imply that compatibility conditions on $\Gamma - R$ line are violated. Zero-dimensional topological invariants at Γ and R must satisfy $N^7_\Gamma = N^4_R$, $N^5_\Gamma = N^6_R$, and $N^5_\Gamma + N^7_\Gamma = 0$ when the Bogoliubov quasiparticle spectrum is fully gapped. The first and second compatibility conditions are violated; thus, two gapless points exist on the $\Gamma - R$ line. In addition, from our classification of gapless points on this line, we find that the gapless points are actually part of a surface node. The analysis for the ^2E pairing is completely the same.

To show that the system is expected to exhibit surface nodes, here we discuss a generator of classifications of Dirac Hamiltonians on Γ-R line. The little co-group of any point in the Γ-R line contains only the three-fold rotation symmetry along the line. Let us start with the case where the two gapless points are in the same position. Then, the Dirac Hamiltonian is given by

$$h_{\boldsymbol{q}} = q_1\sigma_x + q_2\sigma_y + q_3\sigma_z,$$
$$U(C_3) = \mathrm{diag}\left(-1, e^{-i\frac{\pi}{3}}\right),$$

where $\boldsymbol{q} = (q_1, q_2, q_3)$ is the displacement from the gapless points. However, there is no reason why the two gapless points on the line are forced to be in the same position. Indeed, we can add the mass term $M = m\sigma_0$, which shifts the positions of the two gapless points. After computing the energy spectrum of $h_{\boldsymbol{q}} + M$, we obtain $E_{\boldsymbol{q}} = \pm\sqrt{q_1^2 + q_2^2 + q_3^2} + m$. Thus, one can find that the two gapless points are part of a spherical node.

BaPtP with T Pairing

We consider BaPtP with the three-dimensional irreducible representation T. When the pairing symmetry is a multi-dimensional irreducible representation, some symmetries must be broken. As discussed in the main text, the T representation can be decomposed into $B_1 \oplus B_2 \oplus B_3$ of point group D_2 or $A \oplus^1 E \oplus^2 E$ of point group C_3. Here, we assume that the pairing symmetry is B_1 representation of D_2.

First, we have to map the counts of irreducible representations in Eq. (5.3) to those in the lowered space group. After doing that, we obtain the counts of irreducible representations of space group $P2_12_12_1$ as

$$\begin{aligned}&(n_{\Gamma}^{5}, n_{\mathrm{X}}^{2}, n_{\mathrm{X}}^{3}, n_{\mathrm{X}}^{4}, n_{\mathrm{X}}^{5}, n_{S}^{5}, n_{\mathrm{R}}^{2}, n_{\mathrm{R}}^{3}, n_{\mathrm{R}}^{4}, n_{\mathrm{R}}^{5}, n_{U}^{5}, n_{\mathrm{Z}}^{2}, n_{\mathrm{Z}}^{3}, n_{\mathrm{Z}}^{4}, n_{\mathrm{Z}}^{5}, n_{T}^{5}, n_{\mathrm{Y}}^{2}, n_{\mathrm{Y}}^{3}, n_{\mathrm{Y}}^{4}, n_{\mathrm{Y}}^{5})\\ &= (91, 45, 45, 45, 45, 90, 46, 46, 46, 46, 90, 45, 45, 45, 45, 90, 45, 45, 45, 45).\end{aligned} \tag{5.5}$$

Next, based on the weak-pairing assumption, we obtain zero-dimensional topological invariants for B_1 pairing of $P2_12_12_1$ (see Table 5.3):

$$\begin{aligned}&(P_{\Gamma}^{5}, N_{\mathrm{X}}^{2}, N_{\mathrm{X}}^{3}, N_{\mathrm{X}}^{4}, N_{\mathrm{X}}^{5}, N_{\mathrm{R}}^{2}, N_{\mathrm{R}}^{3}, N_{\mathrm{R}}^{4}, N_{\mathrm{R}}^{5}, P_{\mathrm{Z}}^{2}, P_{\mathrm{Z}}^{3}, P_{\mathrm{Z}}^{4}, P_{\mathrm{Z}}^{5}, N_{\mathrm{Y}}^{2}, N_{\mathrm{Y}}^{3}, N_{\mathrm{Y}}^{4}, N_{\mathrm{Y}}^{5})\\ &= (1, 0, 0, 0, 0, 0, 0, 0, 0, 1, 1, 1, 1, 0, 0, 0, 0).\end{aligned} \tag{5.6}$$

There exist the following compatibility conditions among nontrivial zero-dimensional topological invariants in Eq. 5.6:

Table 5.3 EAZ classes at high-symmetry points of magnetic space group $P2_12_12_11'$ with B_1 pairing

Irrep	EAZ class	topological invariant
Γ_5	class BDI	$P_\Gamma^5 = n_\Gamma^5 \mod 2$
X_2	class A	$N_X^2 = n_X^2 - n_X^4$
X_3	class A	$N_X^3 = n_X^3 - n_X^5$
X_4	class A	$N_X^4 = n_X^4 - n_X^2$
X_5	class A	$N_X^5 = n_X^5 - n_X^3$
S_5	class CII	None
R_2	class AII	$N_R^2 = n_R^2 - n_R^4$
R_3	class AII	$N_R^3 = n_R^3 - n_R^5$
R_4	class AII	$N_R^4 = n_R^4 - n_R^2$
R_5	class AII	$N_R^5 = n_R^5 - n_R^3$
U_5	class DIII	None
Z_2	class D	$P_Z^2 = n_Z^2 \mod 2$
Z_3	class D	$P_Z^3 = n_Z^3 \mod 2$
Z_4	class D	$P_Z^4 = n_Z^4 \mod 2$
Z_5	class D	$P_Z^5 = n_Z^5 \mod 2$
U_5	class DIII	None
Y_2	class A	$N_Y^2 = n_Y^2 - n_Y^3$
Y_3	class A	$N_Y^3 = n_Y^3 - n_Y^2$
Y_4	class A	$N_Y^4 = n_Y^4 - n_Y^5$
Y_5	class A	$N_Y^5 = n_Y^5 - n_Y^4$

$$P_\Gamma^5 = P_Z^2 + P_Z^4 \mod 2; \tag{5.7}$$

$$P_\Gamma^5 = P_Z^3 + P_Z^5 \mod 2. \tag{5.8}$$

From Eq. 5.6, we find that both of these compatibility conditions are violated. Thus, the gapless points must exist on Γ-Z line. Furthermore, from our classification of gapless points on this line, we find that the gapless points are actually part of line nodes. The same analysis is applicable to B_2 and B_3 pairings.

BaPtSb with A_1'' Pairing

In the same way as BaPtP, we analyze the topological and nodal nature of BaPtSb with A_1'' pairing. From the first-principles calculation,

$$\begin{aligned}&(n_\Gamma^7, n_\Gamma^8, n_\Gamma^9, n_M^5, n_K^7, n_K^8, n_K^9, n_K^{10}, n_K^{11}, n_K^{12}, n_A^7, n_A^8, n_A^9, n_L^5, n_H^7, n_H^8, n_H^9, n_H^{10}, n_H^{11}, n_H^{12})\\&= (8, 9, 9, 28, 9, 10, 8, 9, 10, 8, 8, 8, 10, 28, 10, 9, 9, 9, 9, 10).\end{aligned} \tag{5.9}$$

Table 5.4 EAZ classes at high-symmetry points of magnetic space group $P\bar{6}m21'$ with A_1'' pairing

Irrep	EAZ class	topological invariant
Γ_7	class BDI	$P_{\Gamma}^{7} = n_{\Gamma}^{7} \mod 2$
Γ_8	class AI	$N_{\Gamma}^{8} = n_{\Gamma}^{8} - n_{\Gamma}^{9}$
Γ_9	class AI	$N_{\Gamma}^{9} = n_{\Gamma}^{9} - n_{\Gamma}^{8}$
M_5	class BDI	$P_{M}^{5} = n_{M}^{5} \mod 2$
K_7	class AI	$N_{K}^{7} = n_{K}^{7} - n_{K}^{8}$
K_8	class AI	$N_{K}^{8} = n_{K}^{8} - n_{K}^{7}$
K_9	class AI	$N_{K}^{9} = n_{K}^{9} - n_{K}^{10}$
K_{10}	class AI	$N_{K}^{10} = n_{K}^{10} - n_{K}^{9}$
K_{11}	class AI	$N_{K}^{11} = n_{K}^{11} - n_{K}^{12}$
K_{12}	class AI	$N_{K}^{12} = n_{K}^{12} - n_{K}^{11}$
A_7	class BDI	$P_{A}^{7} = n_{A}^{7} \mod 2$
A_8	class AI	$N_{A}^{8} = n_{A}^{8} - n_{A}^{9}$
A_9	class AI	$N_{A}^{9} = n_{A}^{9} - n_{A}^{8}$
M_5	class BDI	$P_{L}^{5} = n_{L}^{5} \mod 2$
K_7	class AI	$N_{H}^{7} = n_{H}^{7} - n_{H}^{8}$
K_8	class AI	$N_{H}^{8} = n_{H}^{8} - n_{H}^{7}$
K_9	class AI	$N_{H}^{9} = n_{H}^{9} - n_{H}^{10}$
K_{10}	class AI	$N_{H}^{10} = n_{H}^{10} - n_{H}^{9}$
K_{11}	class AI	$N_{H}^{11} = n_{H}^{11} - n_{H}^{12}$
K_{12}	class AI	$N_{H}^{12} = n_{H}^{12} - n_{H}^{11}$

Based on the above counts of irreducible representations in the normal conducting phase, we obtain $\vec{n}_{\mathrm{SC}}$ as

$$(P_{\Gamma}^{7}, N_{\Gamma}^{8}, N_{\Gamma}^{9}, P_{M}^{5}, N_{K}^{7}, N_{K}^{8}, N_{K}^{9}, N_{K}^{10}, N_{K}^{11}, N_{K}^{12}, P_{A}^{7}, N_{A}^{8}, N_{A}^{9}, P_{L}^{5}, N_{H}^{7}, N_{H}^{8}, N_{H}^{9}, N_{H}^{10}, N_{H}^{11}, N_{H}^{12}) \\ = (0, 0, 0, 0, -1, 1, -1, 1, 2, -2, 0, -2, 2, 0, 1, -1, 0, 0, -1, 1). \tag{5.10}$$

See Table 5.4 for EAZ classes and zero-dimensional topological invariants. These zero-dimensional topological invariants in $\vec{n}_{\mathrm{SC}}$ satisfy all compatibility conditions. This can also be confirmed by the fact $\vec{n}_{\mathrm{SC}} \in \{\mathrm{BS}\}$ as we can expand $\vec{n}_{\mathrm{SC}}$ by

$$\vec{n}_{\mathrm{SC}} = 3\vec{B}_2 - 4\vec{B}_3 - 7\vec{B}_4 - 10\vec{B}_5 + 3\vec{B}_6 + 2\vec{B}_7 - \vec{B}_9 + 5\vec{B}_{10}, \tag{5.11}$$

where $\{\vec{B}_i\}_{i=1}^{10}$ is a basis set of $\{\mathrm{BS}\}$ shown in Eq. (5.14). The coefficients, $\{r_i \mod s_i\}_{i=1}^{10}$, tell us the entry of X_{BdG}, where $\{s_i\}_{i=1}^{10}$ is also defined in Eq. (5.14). Then, we easily find that the entry is $(0, 1, 0, 0, 5, 5) \in X_{\mathrm{BdG}} = (\mathbb{Z}_2)^4 \times (\mathbb{Z}_6)^2$ in our choice of a basis set of $\{\mathrm{BS}\}$. It should be noticed that the entry of X_{BdG} depends on the choice of basis lists of $\{\mathrm{BS}\}$.

As discussed in Refs. [23, 24], the mirror Chern numbers $C_M^{k_z=0,\pi}$ modulo three can be computed only from zero-dimensional topological invariants at high-symmetry momenta. Ref. [8] provides the following formulas

$$C_M^{k_z=0} = N_\Gamma^8 - N_\Gamma^9 + \frac{3}{2}N_\mathrm{K}^7 + \frac{3}{2}N_\mathrm{K}^8 + \frac{1}{2}N_\mathrm{K}^9 - \frac{1}{2}N_\mathrm{K}^{10} - \frac{1}{2}N_\mathrm{K}^{11} + \frac{1}{2}N_\mathrm{K}^{12} \mod 3, \tag{5.12}$$

$$C_M^{k_z=\pi} = N_A^8 - N_A^9 + \frac{3}{2}N_\mathrm{H}^7 + \frac{3}{2}N_\mathrm{H}^8 + \frac{1}{2}N_\mathrm{H}^9 - \frac{1}{2}N_\mathrm{H}^{10} - \frac{1}{2}N_\mathrm{H}^{11} + \frac{1}{2}N_\mathrm{H}^{12} \mod 3. \tag{5.13}$$

Applying the formulas to $\vec{n}_{\mathrm{SC}}$, we find $(C_\mathrm{M}^{k_z=0}, C_\mathrm{M}^{k_z=\pi}) = (0, 1) \mod 3$.

$$\left(\begin{array}{c|cccccccccc}
 & \vec{B}_1 & \vec{B}_2 & \vec{B}_3 & \vec{B}_4 & \vec{B}_5 & \vec{B}_6 & \vec{B}_7 & \vec{B}_8 & \vec{B}_9 & \vec{B}_{10} \\
\hline
s_i & 1 & 1 & 1 & 1 & 2 & 2 & 2 & 2 & 6 & 6 \\
\hline
P_\Gamma^7 & 1 & 0 & 0 & 0 & 0 & 0 & 0 & 1 & 0 & 0 \\
N_\Gamma^8 & 0 & 1 & -4 & 7 & -3 & 0 & 3 & -3 & 1 & -1 \\
N_\Gamma^9 & 0 & -1 & 4 & -7 & 3 & 0 & -3 & 3 & -1 & 1 \\
P_\mathrm{M}^5 & 1 & 1 & 0 & 1 & 0 & 0 & 0 & 1 & 1 & 1 \\
N_\mathrm{K}^7 & 0 & 0 & 2 & -1 & 0 & 0 & 0 & 0 & 0 & 0 \\
N_\mathrm{K}^8 & 0 & 0 & -2 & 1 & 0 & 0 & 0 & 0 & 0 & 0 \\
N_\mathrm{K}^9 & 0 & 1 & -5 & 12 & -6 & 0 & 6 & -6 & 2 & -2 \\
N_\mathrm{K}^{10} & 0 & -1 & 5 & -12 & 6 & 0 & -6 & 6 & -2 & 2 \\
N_\mathrm{K}^{11} & 0 & -1 & 3 & -11 & 6 & 0 & -6 & 6 & -2 & 2 \\
N_\mathrm{K}^{12} & 0 & 1 & -3 & 11 & -6 & 0 & 6 & -6 & 2 & -2 \\
P_\mathrm{A}^7 & 1 & 0 & 0 & 0 & 0 & 0 & 0 & 0 & 0 & 0 \\
N_\mathrm{A}^8 & 0 & 1 & -2 & 3 & -1 & -1 & 3 & 0 & 0 & -1 \\
N_\mathrm{A}^9 & 0 & -1 & 2 & -3 & 1 & 1 & -3 & 0 & 0 & 1 \\
P_\mathrm{L}^5 & 1 & 1 & 0 & 1 & 0 & 0 & 0 & 0 & 0 & 0 \\
N_\mathrm{H}^7 & 0 & 0 & -2 & 7 & -4 & 1 & 3 & -6 & 2 & -1 \\
N_\mathrm{H}^8 & 0 & 0 & 2 & -7 & 4 & -1 & -3 & 6 & -2 & 1 \\
N_\mathrm{H}^9 & 0 & 1 & -1 & 4 & -2 & 0 & 3 & 0 & 0 & -1 \\
N_\mathrm{H}^{10} & 0 & -1 & 1 & -4 & 2 & 0 & -3 & 0 & 0 & 1 \\
N_\mathrm{H}^{11} & 0 & -1 & 3 & -11 & 6 & -1 & -6 & 6 & -2 & 2 \\
N_\mathrm{H}^{12} & 0 & 1 & -3 & 11 & -6 & 1 & 6 & -6 & 2 & -2
\end{array}\right) \tag{5.14}$$

References

1. Zhang T, Jiang Y, Song Z, Huang H, He Y, Fang Z, Weng H, Fang C (2019) Catalogue of topological electronic materials. Nature 566(7745):475–479
2. Tang F, Po HC, Vishwanath A, Wan X (2019) Comprehensive search for topological materials using symmetry indicators. Nature 566(7745):486–489
3. Vergniory MG, Elcoro L, Felser C, Regnault N, Andrei Bernevig B, Wang Z (2019) A complete catalogue of high-quality topological materials. Nature 566(7745):480–485
4. Xu Y, Elcoro L, Song Z-D, Wieder BJ, Vergniory MG, Regnault N, Chen Y, Felser C, Andrei Bernevig B (2020) High-throughput calculations of magnetic topological materials. Nature 586(7831):702–707
5. Tang F, Ono S, Wan X, Watanabe H (2022) High-throughput investigations of topological and nodal superconductors. Phys Rev Lett 129:027001 Jul
6. Sato M, Ando Y (2017) Topological superconductors: a review. Rep Progr Phys 80(7):076501 May
7. Song Z, Zhang T, Fang C (2018) Diagnosis for nonmagnetic topological semimetals in the absence of spin-orbital coupling. Phys Rev X 8:031069 Sep
8. Ono S, Po HC, Watanabe H (2020) Refined symmetry indicators for topological superconductors in all space groups. Sci Adv 6(18):eaaz8367
9. Shang T, Smidman M, Wang A, Chang L-J, Baines C, Lee MK, Nie ZY, Pang GM, Xie W, Jiang WB, Shi M, Medarde M, Shiroka T, Yuan HQ (2020) Simultaneous nodal superconductivity and time-reversal symmetry breaking in the noncentrosymmetric superconductor CaPtAs. Phys Rev Lett 124:207001 May
10. Okamoto Y, Mizutani R, Yamakawa Y, Takatsu H, Kageyama H, Takenaka K (2019) Electronic properties of BaPtP with a noncentrosymmetric cubic crystal structure
11. Hellenbrandt M (2004) The inorganic crystal structure database (ICSD)-present and future. Crystallogr. Rev. 10:17–22
12. Chang G, Xu, S-Y, Wieder BJ, Sanchez DS, Huang S-M, Belopolski I, Chang T-R, Zhang S, Bansil A, Lin H, Zahid Hasan M (2017) Unconventional chiral fermions and large topological fermi arcs in RhSi. Phys Rev Lett 119:206401
13. Ortiz BR, Teicher SML, Kautzsch L, Sarte PM, Ratcliff N, Harter J, Ruff JPC, Seshadri R, Wilson SD (2021) Fermi surface mapping and the nature of charge-density-wave order in the Kagome superconductor CsV_3Sb_5. Phys Rev X 11:041030
14. Ortiz BR, Gomes LC, Morey JR, Winiarski M, Bordelon M, Mangum JS, Oswald IWH, Rodriguez-Rivera JA, Neilson JR, Wilson SD, Ertekin E, McQueen TM, Toberer ES (2019) New kagome prototype materials: discovery of KV_3Sb_5, RbV_3Sb_5, and CsV_3Sb_5. Phys Rev Mater 3:094407
15. Ortiz BR, Teicher SML, Hu Y, Zuo JL, Sarte PM, Schueller EC, Milinda Abeykoon AM, Krogstad MJ, Rosenkranz S, Osborn R, Seshadri R, Balents L, He J, Wilson SD (2020) CsV_3Sb_5: a $\mathbb{Z}_2$ topological Kagome metal with a superconducting ground state. Phys Rev Lett 125:247002
16. Tan H, Liu Y, Wang Z, Yan B (2021) Charge Density Waves and Electronic Properties of Superconducting Kagome Metals. Phys. Rev. Lett. 127:046401 Jul
17. Ono S, Shiozaki K, Watanabe H (2024) Classification of time-reversal symmetric topological superconducting phases for conventional pairing symmetries. Phys Rev B 109:214502. https://doi.org/10.1103/PhysRevB.109.214502
18. Duan W, Nie Z, Luo S, Yu F, Ortiz BR, Yin L, Su H, Du F, An W, Ye C, Lu X, Ying J, Wilson SD, Chen X, Song Y, Yuan H (2021) Nodeless superconductivity in the Kagome metal CsV_3Sb_5. Sci China Phys Mech Astron 64(10):107462
19. Zhao CC, Wang LS, Xia W, Yin QW, Ni JM, Huang YY, Tu CP, Tao ZC, Tu ZJ, Gong CS, Lei HC, Guo YF, Yang XF, Li SY (2021) Nodal superconductivity and superconducting domes in the topological Kagome metal CsV_3Sb_5
20. Kobayashi S, Ai Y, Tanaka Y, Sato M (2019) Majorana multipole response of topological superconductors. Phys Rev Lett 123:097002 Aug

21. He JJ, Tanaka Y, Nagaosa N (2021) . Optical responses of Chiral Majorana edge states in two-dimensional topological superconductors. Phys Rev Lett 126:237002
22. Elcoro L, Bradlyn B, Wang Z, Vergniory MG, Cano J, Felser C, Andrei Bernevig B, Orobengoa D, de la Flor G, Aroyo MI (2017) Double crystallographic groups and their representations on the Bilbao Crystallographic Server. J Appl Crystallogr 50(5):1457–1477
23. Khalaf E, Po HC, Vishwanath A, Watanabe H (2018) Symmetry indicators and anomalous surface states of topological crystalline insulators. Phys Rev X 8:031070
24. Song Z, Zhang T, Fang Z, Fang C (2018) Quantitative mappings between symmetry and topology in solids. Nat Commun 9(1):3530

Chapter 6
Conclusions and Outlook

In this thesis, we developed a symmetry-based approach to topology of superconductors. As seen in Chap. 3, when we tried to generalize symmetry-indicator theory to superconductors, we found that there are two main problems in the original theory: (i) the absence of Pfaffian invariants and (ii) the absence of nontrivial indicators in one dimension. To resolve these problems, we first developed a systematic computation of linear algebra with $\mathbb{Z}_2$-nature, and then we revisited the definition of trivial superconductors. Finally, we generalized the theory of symmetry indicators to superconductors, which enables us to diagnose topological superconducting phases in an efficient way without full information about wave functions. In Chap. 4, we provide a comprehensive understanding of relations between compatibility conditions and superconducting nodes, which is formulated by Atiyah-Hirzebruch spectral sequence. We also developed a classification method of point nodes on arbitrary lines in momentum space. Unifying these two techniques, we proposed a theoretical framework to extensively classify superconducting nodes pinned to lines in momentum space. This framework led to encyclopedic classification tables of Ref. [1] that contain the positions of superconducting nodes in momentum space and their dimensionality such as point and line nodes. These two theories in Chaps. 3 and 4 formed the basis of a systematic diagnosis introduced in Chap. 5. As shown in Chap. 5, we actually applied the diagnostic scheme to several realistic materials with the weak-pairing assumption and provided topological or nodal superconducting nature for each of the possible pairing symmetries.

The framework presented in this thesis opens up various future studies. An obvious direction is the investigation of topological superconductors using our database. Topological flat bands near the Fermi energy open a promising avenue for unconventional superconductivity [2–6]. A recent comprehensive investigation of topological flat bands [6] reveals that many materials in ICSD possess topological flat bands. Since materials in our database largely overlap with them, studying materials listed in these two databases is a promising direction for discovering new topological/nodal superconductors (see Table IV of Supplementary Materials III in Ref. [7]).

S. Ono, *Topology of Superconductors*, Springer Theses,
https://doi.org/10.1007/978-981-97-7399-2_6

It is also an important future study to utilize our results in identifying pairing symmetries. As stated in Chap. 1, our comprehensive investigation of topological and nodal superconductivity could help us reduce the candidates of pairing symmetries in actual superconductors by comparing our results with experimental results. Unfortunately, however, experimental studies on superconducting nature sometimes contradict each other. To resolve the situation, various experimental probes to determine topological and nodal nature should be developed and compared with the outputs of our framework.

Although our diagnostic scheme covers a wide range of topological superconducting phases and superconducting nodes, it is incomplete. That is, our framework misses certain topological superconducting phases and superconducting nodes. For example, while there are many topological superconducting phases with trivial pairing symmetries in the presence of TRS [8, 9], all of them cannot be detected by symmetry indicators. In addition, our symmetry-based approach cannot capture nodes at generic points, as discussed in Sect. 4.4. To establish a complete understanding of the relations between topology and pairing symmetries in all symmetry settings, a more refined method is awaited.

References

1. Ono S, Shiozaki K (2022) Symmetry-based approach to superconducting nodes: unification of compatibility conditions and gapless point classifications. Phys Rev X 12:011021
2. Chatterjee S, Ippoliti M, Zaletel MP (2020) Skyrmion superconductivity: DMRG evidence for a topological route to superconductivity
3. Ma D-S, Xu Y, Chiu CS, Regnault N, Houck AA, Song Z, Andrei Bernevig B (2020) Spin-orbit-induced topological flat bands in line and split graphs of bipartite lattices. Phys Rev Lett 125:266403
4. Khalaf E, Chatterjee S, Bultinck N, Zaletel MP, Vishwanath A (2021) Charged skyrmions and topological origin of superconductivity in magic-angle graphene. Sci Adv 7(19):eabf5299
5. Peri V, Song Z-D, Andrei Bernevig B, Huber SD (2021) Fragile topology and flat-band superconductivity in the strong-coupling regime. Phys Rev Lett 126:027002
6. Regnault N, Xu Y, Li M-R, Ma D-S, Jovanovic M, Yazdani A, Parkin SSP, Felser C, Schoop LM, Phuan Ong N, Cava RJ, Elcoro L, Song Z-D, Andrei Bernevig B (2022) Catalogue of flat-band stoichiometric materials. Nature 603(7903):824–828
7. Tang F, Ono S, Wan X, Watanabe H (2022) High-throughput investigations of topological and nodal superconductors. Phys Rev Lett 129:027001
8. Qin S, Fang C, Zhang F-C, Hu J (2022) Topological superconductivity in an extended s-wave superconductor and its implication to iron-based superconductors. Phys Rev X 12:011030
9. Ono S, Shiozaki K, Watanabe H (2024) Classification of time-reversal symmetric topological superconducting phases for conventional pairing symmetries. Phys Rev B 109:214502. https://doi.org/10.1103/PhysRevB.109.214502

The manufacturer's authorised representative in the EU is Springer Nature Customer Service Centre GmbH, Europaplatz 3, 69115 Heidelberg, Germany. If you have any concerns regarding our products, please contact ProductSafety@springernature.com

Printed and bound by CPI Group (UK) Ltd, Croydon, CR0 4YY
30/09/2026
02225307-0008